MAMMALS AND COUNTRIES OF THE WORLD

A CHECK LIST

GUY COMMEAU

First Edition

Library of Congress Catalog Card Number: TXU 358-746
ISBN 0-9629966-0-2

Cover Design: On the Ball Graphics; Kari Sanford and Rina Dion.

Published by Mammals & Countries of the World Association
        P.O. Box 29
        Clearlake Oaks, CA 95423

Printed in the United States of America
by Griffin Printing

To all the mammals who give so much to the human race.

To my mother and father who gave me the love for nature.

To my wife Louise who shared it with me for over 22 years.

I

MAMMALS AND COUNTRIES OF THE WORLD

Mr. Mrs. Miss:

Address:

Phone:

STARTED:

Write your address, phone and all the numbers of your lists in **PENCIL**.

If you love mammals, travel and photography , the **"MAMMALS AND COUNTRIES OF THE WORLD"** book is for you.

Or the perfect gift for any occasion for someone you know who loves animals.

This book lists 3964 species of mammals.

It can be started anytime by anyone. Get one for your child, write his birth date and keep note when and where HE or SHE saw and recognized the first "DOG",CAT",ELEPHANT" etc.....It is something they will enjoy doing for the rest of their life.

It can also become a very educational HOBBY, as well as a challenge.

This is a book you can enjoy for a lifetime.

You can also keep records of all the places you have been with your " COUNTRIES PROVINCE AND STATE LIST".

Your check list also allows you to keep a record of every photo you take.

The main purpose of this book is to make people more aware of the destruction of habitat, as well as the unnecessary killing of mammals all around the world.

It is my hope that each of you will try to make a better world for all the mammals , and that your CHILDREN'S CHILDREN will be able to have the same list as you have today.

With your **"MAMMALS AND COUNTRIES OF THE WORLD"** book, a visit to the Desert, Mountains, Forest, Lake, River, Ocean or Zoo will never seem the same again.

Since I was very young, mammals have fascinated me so much that I wanted to keep a check list of each one I saw ,(in the wild and in zoos).

Not being able to find anything for such purpose a few years ago, I started what is today this book.

I have to say that, without the Greater Los Angeles Zoo Association docent program, this book would only be a simple check list of mammals.

Most of the descriptions of the principal characteristics that follow each order and family, come from the docent notebook of the Greater Los Angeles Zoo Association; in doing so, this very simple check list became a unique book.

I followed the I.S.I.S. (International Species Inventory System) from U.S. Seal and Dale G.Makey of Minnesota Zoological Garden,St.Paul. This is the Taxonomy book used by United States Zoos.

It is my hope that every owner of this book will try to see as many mammals as possible, in the wild and in zoos, and, if because of the book, some of you visit or, even better, become member of a Zoo, it will be my way to thank them.

Zoos are very important in the world we are living in today, they all need our support.

To all the persons who have helped I am deeply grateful.

Allen Shara lee, Baer Karen, Mrs. William Barton,
Berman Sarah, Bowen Barbara, Brody Jean,
Cammarano Ginny, Clark Herb-Olga, Cook Avis,Deems Alice,
Dr. Maisel Gerald-Laurette, Dr.Wareen Thomas,
Dubois Nancy,Fisher Harvey,Garrett Kimball,Goldstein Vida,
Gomez Louis-Claudine,HallahanJean,Herczog Claudia,
Johnson Sandy, Jones Mickey, King Abigail,
MacMillan June,Maizland Yvonne, Marcondes Paul, Martin Pat,
Meylan Rachel-Bernard, Neidenbach Linda-Larry,
Navojosky Edward, Olitzky Irving,Ott Theresa,
Petrula Dorothea, Rosenfeld Marilyn,Schafer Dale,
Shorr Louise, Stover Adele,Tolstonog Jacques-Jacqueline,
Weeden Tom-Bonnie,

Some with bigger contributions than others, but since all of you were so important from the beginning to the end of this project I thank you all equally.

Errors, great and small, are of course my responsibility.I would appreciate receiving additional informations or corrections which can be included in the next edition .

Guy Commeau
P.O BOX 29
Clearlake Oaks
California 95423

CONTENTS

WHAT IS A MAMMAL ???

Mammals are a very diverse and successful class of vertebrates, occupying a great variety of habitats : Water, Ground, Trees, and Air.

Mammals range in size from the **Pygmy shrew,**which weighs less than 1/6 ounce, to the **Blue whale,**which weighs about 300.000 pounds.

Descendants of reptiles, mammals have evolved many characteristics which allow them much greater success than that of their reptilian ancestors.

CHARACTERISTICS: **Diagnostic - - unique to class mammalia**

MAMMARY GLANDS, Provide milk to nourish young.

HAIR, Present in all mammals at some time during development. Acts as insulation.

DIAPHRAM,Muscular partition separating chest cavity from the abdominal cavity. Movement of diaphram, together with movement of ribs, produces typical breathing action of mammals.

AUDITORY OSSICLES,(ear bones) Mammals possess 3 middle ear bones:
>   1. Malleus (hammer)
>   2.Incus (anvil)
>   3.Stapes (stirrup)

DENTARY, Mandible (lower jaw) consists of single bone on either side, the dentary.

Non - - diagnostic characterictics

ENDOTHERMY, Mammals, like birds,are endothermic. Endothermy permits broad ecological and geographical distribution.

FOUR- CHAMBERED HEART- - DOUBLE CIRCULATORY SYSTEM,Complete separation of oxygenated and deoxygenated blood.

HETERODONT DENTITION, Teeth in most mammals vary in structure and function.Teeth may be specialized for biting, tearing or grinding, permitting a diversity of eating habits.

VIVIPARITY, Most give birth to live young (exception- monotremes).Embryo develops within protected environment of the maternal body.Combination of viviparity and parental care enhances chances of survival in early, critical stages.

ENLARGED BRAIN, Cerebral cortex proportionately larger and more highly developed than in other classes.

HOW TO USE YOUR MAMMALS BOOK

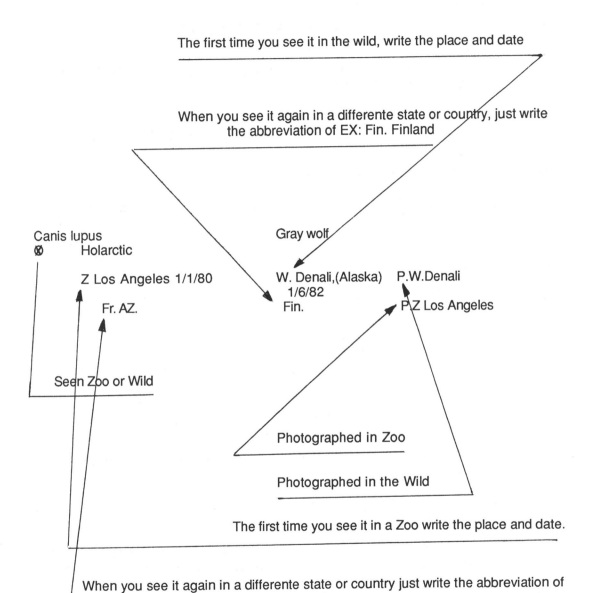

The first time you see it in the wild, write the place and date

When you see it again in a differente state or country, just write the abbreviation of EX: Fin. Finland

Canis lupus
⊗   Holarctic

Gray wolf

Z Los Angeles 1/1/80

W. Denali,(Alaska)   P.W.Denali
1/6/82

Fr. AZ.

Fin.   P Z Los Angeles

Seen Zoo or Wild

Photographed in Zoo

Photographed in the Wild

The first time you see it in a Zoo write the place and date.

When you see it again in a differente state or country just write the abbreviation of it. EX: Fr. France, AZ. Arizona

NOTE: SOME OF THE MAMMALS,ESPECIALLY THE NOCTURNAL ARE VERY DIFFICULT TO SEE, THE SPACE YOU HAVE TO KEEP RECORDS OF VARY WITH THE ABILITY YOU HAVE TO SEE THEM.

MAMMALS LIFE LIST: Is the total of mammals you saw in the Wild, Domestic, Zoo.

MAMMALS WILD LIST: Is all the mammals you saw in the Wild.

MAMMALS ZOO LIST: Is all the mammals you saw in Zoos.

MAMMALS WILD LIST OF NORTH AMERICA: Is all the mammals you saw in the wild in North America. (North America is United States and Canada).

MAMMALS ZOO LIST OF NORTH AMERICA: Is all the mammals you saw in zoos in North America.

MAMMALS PHOTOGRAPHS LIST: Is all the mammals you photographs. You can also keep a seperate list for Wild and Zoo mammals.

MY MAMMALS LIFE LIST IS: _____

MY MAMMALS WILD LIST IS: _____

MY MAMMALS ZOO LIST IS: _____

MY MAMMALS WILD LIST OF NORTH AMERICA IS: _____

MY MAMMALS ZOO LIST OF NORTH AMERICA IS: _____

MY MAMMALS PHOTOGRAPHS LIST IS: _____

WILD:_____

ZOO:_____

MY STATE LIST IS: _____

MY CANADA PROVINCE LIST IS: _____

MY COUNTRIES LIST IS: _____

MY LIST FOR EACH ORDER IS

MONOTREMATA _____

MARSUPIALIA _____

INSECTIVORA _____

DERMOPTERA _____

CHIROPTERA _____

PRIMATES _____

EDENTATA _____

PHOLIDOTA _____

LAGOMORPHA _____

RODENTIA _____

CETACEA _____

CARNIVORA _____

PINNIPEDIA _____

TUBULIDENTATA _____

PROBOSCIDEA _____

HYRACOIDEA _____

SIRENIA _____

PERISSODACTYLA _____

ARTIODACTYLA _____

CLASSIFICATION

In order to study the thousands of forms of life in a systematic manner, scientists have classified and arranged them into groups based on their natural relationships.

Example of a classification of the WOLF

KINGDOM	Animal
PHYLUM	Chordata
SUBPHYLUM	Vertebrata
CLASS	Mammalia
ORDER	Carnivora
FAMILY	Canidae
GENUS	Canis
SPECIES	Lupus

IN THIS BOOK **CLASS MAMMALIA** IS DIVIDED INTO **19** ORDERS.

THE **19 ORDERS** ARE DIVIDED INTO 118 **FAMILIES**

AND A TOTAL OF **3964 SPECIES**

ORDER	ABBREVIATION	FAMILY	SPECIES
MONOTREMATA	MO.	2	3
MARSUPIALIA	MA.	12	234
INSECTIVORA	IN.	9	380
DERMOPTERA	DE.	1	2
CHIROPTERA	CH.	16	855
PRIMATES	PR.	10	179
EDENTATA	ED.	3	29
PHOLIDOTA	PH.	1	7
LAGOMORPHA	LA.	2	59
RODENTIA	RO.	28	1643
CETACEA	CE.	8	79
CARNIVORA	CA.	7	245
PINNIPEDIA	PI.	2	34
TUBULIDENTATA	TU.	1	1
PROBOSCIDAE	PRO.	1	2
HYRACOIDEA	HY.	1	7
SIRENIA	SI.	2	4
PERISSODACTYLA	.PE	3	16
ARTIODACTYLA	AR.	9	185

THE NEXT LIST IS THE COMMON MAMMAL'S NAME OF THE 19 ORDERS

This will help you learn that
KOALA, WOMBATS, KANGOROOS,
belong to the order MARSUPIALIA

LEMURS, MARMOSET, MAN
belong to the order PRIMATES

DOGS, RACCOONS, HYENAS, CATS
belong to the order CARNIVORA etc,etc...

MONOTREMATA: Echidnas, Platypus

MARSUPIALIA: New world opossums, Marsupial' mice, rats',Marsupial'mole', Bandicoots, South american'rat' opossums,Possums and Cuscuses, Gliders' striped possums'ringtails,Pygmy possums, Honey possum, Koala, Wombats, Kangaroos, Wallaroos, Wallabies, Rat kangaroos.

INSECTIVORA: Solenodons, Tenrecs, Otter shrews, Golden moles, Hedgehogs,Moles, Shrews, Elephant shrews, Tree shrews.

DERMOPTERA: Flying lemurs.

CHIROPTERA: Bats.

PRIMATES: Lemurs, Woolly indris, Sifaka, Indris, Aye-aye, Lorises, Potos, Galagos,Tarsiers, Marmosets, Tamarins, New world monkeys, Old world monkeys, Gibbons, Siamangs, Great apes, Man.

EDENTATA: Anteaters, Sloths, Armadillos.

PHOLIDOTA: Pangolins.

LAGOMORPHA: Pikas, Hares, Rabbits.

RODENTIA: Mountain beaver, Squirrels, Chipmunks, Marmots, Prairie dogs, Pocket gophers, Kangaroo rats, Pocket mice, Beavers, Scaly tailed squirrel, Springhare, Deer mice, Woodrats, Lemmings, Voles, Muskrats, Gerbils, Hamsters, Old world rats & mice, Dormice, Jumping mice,Jerboas, Old world porcupines, New world porcupines, Guinea pig, Cavy, Capybara, Agoutis, Pacas, Chinchillas, Viscachas, Hutias, Nutria, Octodonts, Tuco tucos, Spiny rats, Dasie rat, African mole rats, Blesmols, Gundis.

CETACEA: River dolphins, The oceanic dolphins, Porpoises, Pilot whales, Killer whale, False killer whale, Beluga, Narwhal, Sperm whales, Beaked whales, Gray whale, Rorquals, Right whales.

CARNIVORA: Dogs, Wolves, Coyotes, Jackals, Foxes, Bears, Raccoons and relatives, Weasels, Badgers, Skunks, Otters, Civets, Genets, Mongooses, Hyenas, Aardwolf, Cats, Lion, Tigers, Cheetah, Leopards.

THE NEXT LIST IS THE COMMON MAMMAL'S NAME OF THE 19 ORDERS

PINNIPEDIA: Sea lions, Fur seals, Walrus, True seals.

TUBULIDENTATA: Aardvark

PROBOSCIDEA: Elephants.

HYRACOIDEA: Hyraxes.

SIRENIA: Dugong, Manatees.

PERISSODACTYLA: Horses, Zebras, Asses, Tapirs, Rhinoceros.

ARTIODACTYLA: Pigs, Peccaries, Hippopotamuses,Camels, Guanaco, LLama, Alpaca, Vicuna, Mouse deer or Chevrotains, Deer, Elk, Moose, Giraffes, Okapi, Pronghorn, Antelope, Cattle, Buffalo, Goats, Sheep.

An introduction to the principals characteristics will follow each ORDER.From the well known, Example:Order CARNIVORA (Cats,dogs,etc..)."Carnivores are primarily hunters with a good sense of smell".To the lesser known,Example:Order MONOTREMATA (Echidnas and Platypus) of Australia and New Guinea, "The only egg laying mammals"

You will also learn facts following each FAMILY.Example: Family EQUIDAE (Horses,Zebras,Asses)Equidae developed originally in North America, but for some unknown reason,horses disappeared from the New World before historic times.Family LEPORIDAE (Hares and Rabbits) Hares live in open areas,the young are born furred,eyes open and ready to move soon after birth.In contrast Rabbits occupy burrows and give birth to young blind and naked.

RED DATA BOOK CATEGORIES

E= (endangered) taxa in danger of extinction, and whose survival is unlikely if the causal factors continue operating. Included are taxa whose numbers have been reduced to a critical level, or whose habitats have been so drastically reduced that they are deemed to be in immediate danger of extinction.Also included are taxa that are possibly already extinct, but have definitely been seen in the wild in past 50 years.

V= (vulnerable) taxa believed likely to move into "endangered"category in the future if the casual factors continue operating. Included are taxa of which most or all the populations are decreasing because of over exploitation,extensive destruction of habitat,or other environmental disturbance;taxa with populations that have been seriously depleted and whose ultimate security has not yet been assured;and taxa with populations that are still abundant but are under threat from severe adverse factors throughout their range.

R= (rare) taxa with small world populations that are not at present "endangered" or "vulnerable" but are at risk. These taxa are usually localised within restricted geographical areas or are thinly scattered over a more extensive range.

I= (indeterminate) taxa known to be "endangered" "vulnerable" or "rare" but where there is not enough information to say which of the three categories is appropriate.

ENDANGERED SPECIES ACT OF 1973 (U.S.A.)

E*= (endangered) any species which is in danger of extinction throughout all or a significant portion of its range.

T*= (threatened) any species which is likely to become an endangered species within the foreseeable future throughout all or a significant portion of its range.

HOW TO KEEP RECORDS OF UNITED STATES
CANADA AND COUNTRIES CHECK LIST

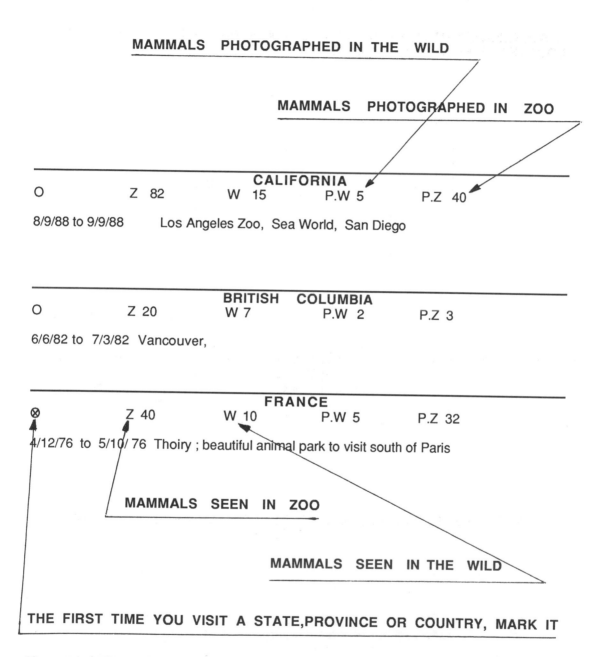

MAMMALS PHOTOGRAPHED IN THE WILD

MAMMALS PHOTOGRAPHED IN ZOO

CALIFORNIA

O   Z 82   W 15   P.W 5   P.Z 40

8/9/88 to 9/9/88   Los Angeles Zoo, Sea World, San Diego

BRITISH COLUMBIA

O   Z 20   W 7   P.W 2   P.Z 3

6/6/82 to 7/3/82 Vancouver,

FRANCE

⊗   Z 40   W 10   P.W 5   P.Z 32

4/12/76 to 5/10/ 76 Thoiry ; beautiful animal park to visit south of Paris

MAMMALS SEEN IN ZOO

MAMMALS SEEN IN THE WILD

THE FIRST TIME YOU VISIT A STATE,PROVINCE OR COUNTRY, MARK IT

You can keep notes of the year's visit, special mammals, special events etc.....

HOW TO KEEP RECORDS OF YOUR SLIDES

A SYSTEMATIC METHOD IS VIRTUALLY A MUST WHEN YOU HAVE TO LOCATED A SLIDE AT A MOMENT'S NOTICE.

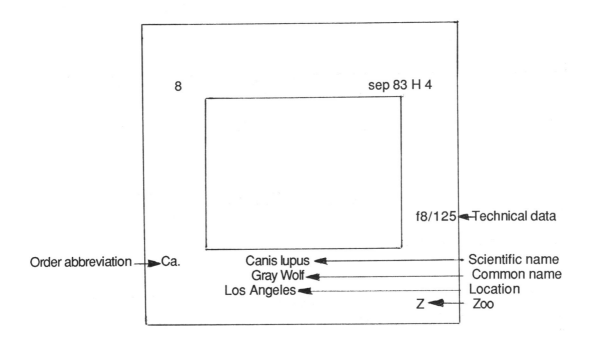

YOU CAN KEEP YOUR MAMMAL SLIDES IN A DOUBLE DECKER SLIDE1500, ALSO BY ORDER

ORDERS AND FAMILIES

	Order	Family	Page
Echidna	Monotremata	Tachyglossidae	2
Platypus	Monotremata	Ornithorhynchida	2
New world Opossums	Marsupialia	Didelphidae	4 - 7
Marsupial (Mice,Rats, Carnivores)	Marsupialia	Dasyuridae	8 - 10
Marsupial mole	Marsupialia	Notoryctidae	1 0
Bandicoots	Marsupialia	Permelidae	1 1
South American Rat Opossums	Marsupialia	Caenolestidae	1 2
Possums,Cuscuses	Marsupialia	Phalangeridae	1 2 - 1 3
Gliders,Ringtails, Striped possums	Marsupialia	Petauridae	1 3 - 1 4
Pygmy possums	Marsupialia	Burramyidae	1 5
Honey possum	Marsupialia	Tarsipedidae	1 5
Koala	Marsupialia	Phascolarctidae	1 5
Wombats	Marsupialia	Vombatidae	1 6
Kangoroos, Wallaroos, Wallabies,Rat Kangaroos	Marsupialia	Macropodidae	1 6 - 2 0
Solenodons	Insectivora	Solenodontidae	2 4
Tenrecs	Insectivora	Tenrecidae	2 4 - 2 5
Otter shrews	Insectivora	Potamogalidae	2 6
Golden moles	Insectivora	Chrysochloridae	2 6 - 2 7
Hedgehogs	Insectivora	Erinaceidae	2 7 - 2 8
Moles	Insectivora	Talpidae	2 8 - 2 9
Shrews	Insectivora	Soricidae	2 9 - 4 4
Elephant shrews	Insectivora	Macroscelididae	4 4 - 4 5
Tree shrews	Insectivora	Tupaiidae	4 5 - 4 6
Flying lemurs	Dermoptera	Cynocephalidae	5 0
Flying foxes,Old world fruit bats	Chiroptera	Pteropodidae	5 2 - 5 9
Mouse tailed bats	Chiroptera	Rhinopomatidae	5 9
Sheath tailed bats	Chiroptera	Emballonuridae	5 9 - 6 2
Bulldog bats	Chiroptera	Noctiliondae	6 2
Slit faced bats	Chiroptera	Nycteridae	6 2 - 6 3
False vampire bats	Chiroptera	Megadermatidae	6 3
Horseshoe and Leaf nosed bats	Chiroptera	Rhimolophidae	6 3 - 7 0

ORDERS AND FAMILIES

ORDERS AND FAMILIES

ORDERS AND FAMILIES

ORDER:MONOTREMATA

Monotremata are the only egg-laying mammals.

They lay leathery, shell-covered, large-yolked eggs, incubated by the female and hatched outside the maternal body. They are toothless (except young platypus) with an elongated, beak-like rostrum. They are confined to the areas of Australia, New Guinea and Tasmania. No fossil records of the order have been found outside the Australian region.This primitive order is an evolutionary dead end and is not ancestral to other living mammals.

Monotremes have both mammalian and reptilian characteristics.

Some mammalian characteristics;

1.   Hair (echidna has hair as well as spines which are also a form of keratin)

2.   Mammary glands (young suckle from 2 haired areas on abdominal wall, nipples or teats not present).

3.   Heterothermic (constant body temperature maintained under favorable conditions or it may fluctuate with the ambient environmental temperature)

Some reptilian characteristics:

1.   Lay eggs (egg tooth and caruncle in hatchling).

2.   Cloaca.

3.   Coracoid bones present in shoulder girdle attachment, giving animal reptilian stance.

FAMILY : TACHYGLOSSIDAE, Echidnas

These animals have robust bodies, covered with a mixture of hair and sturdy spines The clawed feet are broad and powerful. Echidnas are powerful diggers escaping predators by burrowing or wedging themselves tightly into crevices. The long sticky tongue aids the echidna in obtaining its food which consists of ants, termites and worms. Usually one leathery shelled egg is laid and incubated in the temporary pouch on the mother's abdomen. This family contains 2 genera; Tachyglossus of Australia and Zaglossus of New Guinea. Tachyglossus is a true hibernator and becomes torpid in response to cold and lack of food.

Tachyglossus aculeatus   Short nosed Echidna
O Australia, New Guinea, Tasmanian
 Z   W   P

Zaglossus bruijni   Long nosed Echidna
O New Guinea
 Z   W   P

FAMILY: ORNITHORHYNCHIDAE, Platypus

The body of the platypus is flattened and elongated with a beaver-like tail which acts as a paddle. The limbs are short and stout; the feet are webbed and clawed; the fur is dense with woolly underfur. The snout is elongated with a flattened, leathery bill which has good tactile ability. The eye and ear openings lie in a furrow that is closed by folds of skin when the animal is submerged. The platypus is semi-aquatic, burrowing into muddy stream banks and feeding along stream bottoms on small crustaceans, worms, tadpoles and small fish. The females dig burrows into which usually 2 eggs are laid. The eggs are incubated by the female wrapping herself around them; they hatch in about 10 days. The young suckle about 5 months.

Ornithorhynchus anatinus Platypus
O Australia
 Z   W   P

TOTAL MONOTREMATA: O   Z   W   P

ORDER:MARSUPIALIA

Marsupials form a separate evolutionary line from other mammals.

They have been distinct from placental mammals for at least 135 million years.

They are relatively primitive mammals, especially in the development of the brain. Once world-wide in distribution, they were unable to compete successfully with the more advanced placentals and are now confined to the Australian region and a few areas of the Americas. Isolated in an insular situation, they have undergone adaptive radiation and occupy nearly all types of ecological niches in Australia. There are carnivorous, herbivorous, fossorial, terrestrial and arboreal forms. In fact, most marsupials have functional counterparts among the placental mammals. Marsupials have a short gestation period; the young born in an extremely immature condition with only the forelimbs, mouth, and certain parts of the nervous system being well developed. Most marsupials have some type of pouch or fold in which the teats are located. At birth, the infant makes its way into the marsupium and attaches itself to a teat or nipple. It remains there usually until it approximates the size and development of a comparable placental young at birth. Most "typical" pouches open forward, burrowers have pouches which open to the rear, some have merely a temporary fold of skin, others no pouch at all

FAMILY: DIDELPHIDAE, New World Opossums

This is the oldest known family of marsupials whose members are primitive and generalized. The big toe is clawless and opposable. Some members of the family have a pouch; in some it is absent; and still in others it is 2 folds of skin. There are 50 rooted and sharp teeth. Opossums are mostly arboreal and nocturnal. They are omnivorous. "Playing possum" is a term for feigning death, a trait of the North American Opossum (Didelphis virginiana). It is a reaction controlled by the nervous system.

Caluromys derbianus
O   S. Mexico to W. Ecuador

Derbian opossum

Caluromys lanatus
O   Colombia to Paraguay, Brazil

Wooly opossum

Caluromys philander
O   Venezuela to Brazil

Philander opossum

Caluromysiops irrupta
O   Peru

Black-shouldered opossum

Chironectes minimus
O   Mexico to Argentina

Yapok

Didelphis albiventris
O   Venezuela to Argentina

Azaras opossum

Didelphis marsupialis
O   Central and South America

Common opossum

Didelphis virginiana
O   Canada to Costa Rica
    Z   W   P.W
                                                         P.Z
Dromiciops australis   Monitos del monte
O   Chile, Argentina.

Glironia criniger
O   Amazonian Peru

Bushy-tailed opossum

Glironia venusta
O   Ecuador to Bolivia

Bushy-tailed opossum

Lestodelphis halli
O   S.Patagonia

Patagonian opossum

Lutreolina crassicaudata
O Argentina to Paraguay

Little water opossum

Marmosa aceramarcae
O   Bolivia

Murine opossum

Marmosa agilis
O Brazil, Paraguay, Argentina to Bolivia

Murine opossum

4

Marmosa agricolai   Murine opossum
O Brazil (Ceara)

Marmosa canescens   Grayish mouse opossum
O   Mexico

Marmosa cinerea   Ashy opossum
O   Venezuela to Paraguay

Marmosa constantiae   Murine opossum
O   Argentina, Bolovia

Marmosa domina   Murine opossum
O   Amazonian Brazil

Marmosa dryas   Murine opossum
O   Venezuela

Marmosa elegans   Murine opossum
O Chile, Argentina, Bolivia

Marmosa emiliae   Murine opossum
O Paraguay, Brazil,Guyana

Marmosa formosa   Murine opossum
O N. Argentina

Marmosa fuscata   Opossum
O Venezuela, Colombia, Trinidad

Marmosa germana   Opossum
O Peru, Ecuador

Marmosa grisea   Opossum
O Paraguay

Marmosa impavida   Opossum
O Panama to Peru

Marmosa incana   Opossum
O Brazil

Marmosa invicta   Panama mouse opossum
O   Panama

Marmosa karimii   Opossum
O   Brazil (Mato Grosso)

Marmosa lepida   Opossum
O   Ecuador, Bolivia to Surinam

5

Marmosa leucastra   Opossum
O   Peru

Marmosa mapiriensis   Opossum
O   Bolivia, Peru

Marmosa marica   Opossum
O   Venezuela

Marmosa mexicana   Mexican mouse opossum
O   Mexico to Costa Rica

Marmosa microtarsus   Opossum
O   Brazil

Marmosa murina   Murine opossum
O   Tobago to Venezuela to Brazil

Marmosa noctivaga   Opossum
O   Brazil to Ecuador,Bolivia

Marmosas ocellata   Opossum
O   Bolivia

Marmosa parvidens   Opossum
O   Guyana

Marmosa phaea   Opossum
O   Colombia, Ecuador

Marmosa pusilla   Opossum
O   Argentina, Bolivia, Paraguay

Marmosa quichua   Opossum
O   Peru

Marmosa rapposa   Opossum
O   Peru, Bolivia

Marmosa regina   Opossum
O   Colombia

Marmosa robinsoni   Opossum
O   Panama to Ecuador, Trinidad

Marmosa rubra   Opossum
O   Ecuador, Peru

Marmosa scapulata   Opossum
O   Brazil

Marmosa tatei   Opossum
O   Peru

Marmosa tyleriana   Opossum
O   Venezuela

Marmosa unduaviensis   Opossum
O   Bolivia

Marmosa velutina   Opossum
O   Argentina, Brazil

Marmosa yungasensis   Opossum
O   Bolivia

Metachirus nudicaudatus   Brown four-eyed opossum
O   Nicaragua to Argentina

Monodelphis adusta   Short-tailed opossum
O   Panama to Peru

Monodelphis americana   Opossum
O   Brazil

Monodelphis brevicaudata   Opossum
O   The Guianas,Brazil, Venezuela

Monodelphis dimidiata   Opossum
O   Uruguay, Brazil

Monodelphis domestica   Opossum
O   Brazil,Bolivia,Paraguay

Monodelphis fosteri   Opossum
O   Argentina

Monodelphis henseli   Opossum
O   Brazil, Paraguay, Argentina

Monodelphis osgoodi   Short-tailed opossum
O   Bolivia, Peru

Monodelphis scalops   Opossum
O   Brazil

Monodelphis sorex   Opossum
O   Brazil

Monodelphis unistrata   Opossum
O   Sao Paulo,(Brazil)

Philander Mc. ilhennyi   Four-eyed opossum
O   Peru

Philander opossum   Gray four-eyed opossum
O   Central and South America

7

FAMILY:DASYURIDAE Marsupial "Mice", "Rats", "Carnivores"

This very diverse family includes the Tasmanian devil, native "Cats" and marsupial "Mice"The family is closely related to the Didelphidae both of which are similar to the primitive marsupial stock.The teeth are adapted for an insectivorous or carnivorous diet; the canines are large and sharp; the molars have 3 sharp cusps. Most members of the family are nocturnal predators; some terrestrial, a few arboreal.The pouch is often absent; when present, it is often poorly developed and opens posteriorly.The family ranges in size from that of a shrew to a medium sized dog.

Dasycercus cristicauda   Crest-tailed marsupial mouse
O   Australia

Dasyuroides byrnei   Kowari
O   Nt.

Dasyurus albopunctatus   New Guinea dasyure
O   New Guinea

Dasyurus geoffroii   Chuditch
O   Australia

Dasyurus hallucatus   Satanellus
O   Wa, Nt, Qld

Dasyurus maculatus   Tiger cat
O   Qld, Nsw, Vic, Sa, Tasmania

Dasyurus Viverrinus   Quoll   (E*)
O   Nsw, Vic, Sa, Tasmania

Myoictis melas   Three-striped dasyure
O   New Guinea, Aru Isl.

Neophascogale lorentzii   Neophascogale
O   New Guinea

Phascolosorex doriae   Sorecine dasyure
O   New Guinea

Phascolosorex dorsalis   Sorecine dasyure
O   New Guinea

Sarcophilus harrisii   Tasmanian devil
O   Tasmania

Myrmecobius fasciatus   Numbat (E*)
O   S.W. West Australia

Antechinomys laniger   Kultarr (E)
O   Vic. Nsw. Qld

Antechinomys spenceri   Wuhl-wuhl
O   Wa. Sa. Nt.

Antechinus apicalis
O   Wa.

Dibbler (I)

Antechinus bellus
O   Nt.

Fawn marsupial mouse

Antechinus flavipes
O   E. Australia

Yellow-footed marsupial mouse

Antechinus godmani
O   Qld.

Godmans marsupial mouse

Antechinus mac donnellensis
O   Wa. Nt.

Fat-tailed marsupial

Antechinus maculatus
O   Australia

Pygmy marsupial mouse

Antechinus melanurus
O   New Guinea

Marsupial mouse

Antechinus minimus
O   Sa. Vic. Tasmania

Swamp antechinus

Antechinus naso
O   New Guinea

Marsupial mouse

Antechinus rosamondae
O   Wa.

Little red antechinus

Antechinus stuartii
O   Qld. Nsw. Vic.

Brown antechinus

Antechinus swainsonii
O   Nsw. Vic. Sa. Tasmania

Dusky marsupial mouse

Antechinus wilhemlina
O   New Guinea

Marsupial mouse

Murexia longicaudata
O   New Guinea

Long-tailed marsupial mouse

Phascogale calura
O   Australia

Red-tailed wambenger (I)

Phascogale tapoatafa
O   Australia

Brush-tailed phascogale

Planigale ingrami
O   Qld. Nt. Wa.

Northern planigale (E*)

Planigale novaeguineae
O   New Guinea

New Guinea planigale

9

MARSUPIAL "MICE","RATS", "CARNIVORES"

Planigale subtilissima   Kimberley planigale
O   Wa.

Planigale teniurostris   Narrow-nosed planigale (E*)
O   Nsw. Qld. Wa

Sminthopsis crassicaudata   Fat-tailed dunnart
O   Australia

Sminthopsis froggatti   Stripe-faced dunnart
O   Sa. Qld. Nt. Wa.

Sminthopsis granulipes   Granule-footed sminthopsis
O   Wa.

Sminthopsis hirtipes   Hairy-footed dunnart
O   Australia

Sminthopsis leucopis   White-footed dunnart
O   Nsw. Vic. Tasmania

Sminthopsis longicaudata   Long-tailed dunnart (E*)
O   Wa. Cen. Australia

Sminthopsis macroura   Darling downs dunnart
O   Qld. Nsw.

Sminthopsis murina   Common dunnart
O   Australia

Sminthopsis nitela   Daly river dunnart
O   Nt, Kimberley

Sminthopsis psammophila   Sandhill dunnart (E*)
O   Nt.

Sminthopsis rufigenis   Red-cheeked dunnart
O   Qld.

FAMILY: NOTORYCTIDAE,   Marsupial   " Mole "

Marsupial "Mole" have many adptations for fossorial (burrowing) life. The eyes are vestigial and non-functional. There is a horny shield on the nose and the nostrils are narrow slits. The claws of digits 3 and 4 of the forefoot are greatly enlarged and act as a spade for digging. The central 3 digits of the hind foot are enlarged. The marsupial " Mole" is similar in habits to the placental mole, an example of convergent evolution. They differ in habitat preference in that the marsupial "Mole" inhabits dry regions whereas the placental moles characteristically inhabit moist soil. The food is mainly invertebrate larvae

Notoryctes typhlops   Marsupial mole
O   Australia

Bandicoots have an elongated nose used for rooting in soil. The hind legs are specialized, the 4 digit is elongated as an adaptation for running. The second and third digits which are partially fused (syndactylous) are used in grooming. Bandicoots are terrestrial and mainly nocturnal. They have insectivore-like dentition for a mainly insectivorous diet . The pouch opens to the rear.

Echymipera clara   Spiny bandicoot (R)
O   New Guinea

Echymipera kalubu   Bandicoot
O   New Guinea, Bismarck Arch.

Echymipera rufescens   Spiny bandicoot
O   Cape York, New Guinea

Isoodon auratus   Golden bandicoot
O   C. Australia

Isoodon macrourus   Brindled bandicoot
O   Nsw. Qld.

Isoodon obesulus   Short-nosed bandicoot
O   Australia, Tasmania

Macrotis lagotis   Rabbit bandicoot (E)
O   Australia

Microperoryctes murina   Murine bandicoot
O   West Irian, New Guinea

Perameles bougainville   Barred bandicoot (R)
O   Australia

Perameles gunnii   Tasmanian barred bandicoot
O   Vic. Tasmania

Perameles nasuta   Long-nosed bandicoot
O   Qld. Nsw. Vic.

Peroryctes longicaudata   New Guinea bandicoot
O   New Guinea

Peroryctes papuensis   Papuan bandicoot
O   New Guinea

Peroryctes raffrayanus   New Guinea bandicoot
O   New Guinea

Rhynchomeles prattorum   Ceram Island bandicoot
O   Ceram isl.

FAMILY:CAENOLESTIDAE, South American " Rat " Opossums

Rat or mouse sized marsupials which resemble shrews in appearance, having an elongated head and small eyes. They have a well developed sense of smell and eat mostly insects. They are restricted to forested areas of the Andes mountains.

Caenolestes caniventer   Rat opossum
O   Ecuador

Caenolestes convelatus   Rat opossum
O   N.Ecuador

Caenolestes fuliginosus   Rat opossum
O   N. Ecuador (Andes)

Caenolestes obscurus   Rat opossum
O   Colombia, Venezuela

Caenolestes tatei   Rat opossum
O   N. Ecuador

Lestoros inca   Peruvian rat opossum
O   S. Peru

Rhyncholestes raphanurus   Chilean rat opossum
O   S. Chile , Chiloe

FAMILY: PHALANGERIDAE, Possums and Cuscuses

These moderate sized marsupials are primarily arboreal with large hands and feet and a prehensile tail. Most inhabit wooded areas. Phalangerids are omnivorous, eating a wide variety of plants as well as insects, eggs and young birds. The brush-tailed possum, a member of this family, is very adaptable, frequenting suburban areas and feeding on cultivated plants.

Phalanger atrimaculatus   Cuscus (R)
O   New Guinea

Phalanger celebensis   Celebes cuscus
O   Celebes

Phalanger gymnotis   Cuscus
O   New Guinea

Phalanger maculatus   Spotted cuscus (R)
O   Qld. New Guinea

Phalanger orientalis   Gray cuscus
O   Qld. New Guinea

Phalanger ursinus   Bear phalanger
O   Celebes

Phalanger vestitus
O   New Guinea

Silky cuscus

Trichosurus arnhemensis
O   Barrow Isl. Nt

Northern brush possum

Trichosurus caninus
O   Qld. Nsw. Vic.

Short-eared possum

Trichosurus vulpecula
O   Australia

Brush-tailed possum

Wyulda squamicaudata
O   Kimberley of North Wa.

Scaly-tailed possum (E*)

FAMILY: PETAURIDAE, Gliders," Striped possums", Ringtails

Petaurids are rather small, some with prehensile tails, others with long bushy tails. They are nocturnal, arboreal creatures inhabiting wooded areas, Ringtails are strictly herbivorous. Some of the "striped possums" are insectivorous, using an elongated 4 th. digit and tongue to extract insects from trees. Their incisors are used to tear away the tree bark. As in skunks, the striped pattern is associated with a powerful musky scent. Gliders have furred membranes that extend from the body between the limbs and function as lifting surfaces for gliding. The gliding style is similar to that of the flying squirrels. The greater glider has a specialized diet feeding mainly on eucalyptus leaves and blossoms. Sugar gliders live in family groups and rely on scent marking for individual recognition and group organization.

Dactylonax palpator
O   New Guinea

Long-fingered possum

Dactylopsila megalura
O   New Guinea

Striped possum

Dactylopsila tatei
O   New Guinea

Striped possum

Dactylopsila trivirgata
O   Qld. New Guinea

Striped possum

Distoechurus pennetus
O   New Guinea, Papua

Pen-tailed phalanger

Gymnobelideus leadbeateri
O   Vic.

Leadbeaters possum (E)

Hemibelideus lemuroides
O   Qld.

Brush-tipped ringtail

Petaurus australis
O   Qld. Nsw. Vic.

Yellow-bellied glider

13

Petaurus breviceps   Sugar glider
O   Australia,Tasmania, New Guinea

Petaurus norfolcensis   Squirrel glider
O   Vic. Nsw. Qld.

Petropseudes dahli   Rock-haunting ringtail
O   N.Wa. N.Nt.

Pseudocheirus albertisi   Phalanger
O   New Guinea

Pseudocheirus archeri   Striped ringtail
O   Qld.

Pseudocheirus canescens   Phalanger
O   New Guinea

Pseudocheirus caroli   Phalanger
O   New Guinea

Pseudocheirus corinnae   Phalanger
O   New Guinea

Pseudocheirus cupreus   Phalanger
O   New Guinea

Pseudocheirus forbesi   Phalanger
O   New Guinea

Pseudocheirus herbertensis   Herbert river ringtail
O   N.E.Qld.

Pseudocheirus mayeri   Phalanger
O   New Guinea

Pseudocheirus peregrinus   Common ringtail
O   W.New Guinea

Pseudocheirus pygmaeus   Ringtail
O   Australia, Tasmania

Pseudocheirus schlegeli   Phalanger
O   New Guinea

Schoinobates volans   Greater glider
O   E.Australia

FAMILY : BURRAMYIDAE, Pygmy Possums

These small marsupials have large eyes, mouse-like ears and a long prehensile tail. They live in wooded areas and are apparently insectivorous-omnivorous, though feeding habits of some are not known. Some become torpid during cold weather and in 2 genera, the tail becomes enlarged with fat as winter approaches

Acrobates pygmaeus   Pygmy glider
O   Qld. to Sa. Nsw.

Burramys parvus   Mountain pygmy possum (E*)
O   Vic.

Cercartetus caudatus   Long-tailed pygmy possum
O   Qld. New Guinea

Cercartetus concinnus   Mundarda
O   Wa. Sa. Vic.

Cercartetus lepidus   Tasmanian pygmy possum
O   Sa. Tasmania

Cercartetus nanus   Pygmy possum
O   Sa. Vic. Nsw. Tasmania

FAMILY:TARSIPEDIDAE, Honey Possum

The honey possum is a very small, slender-nosed marsupial with a long prehensile tail. It is highly specialized; its relationship to other marsupials is obscure. The cheek teeth are degenerate, only the upper canines and 2 lower incisors are well developed. The long tongue has bristles at the tip similar to that of some nectar-feeding bats. It is used to probe into flowers for nectar, pollen and small insects. The honey possum lives in forested and shrub grown areas

Tarsipes spencerae   Honey possum
O   S.W. Australia

FAMILY : PHASCOLARCTIDAE, Koala

The koala has a chunky, tailless body with tufted ears and a naked nose. It is highly specialized, feeding on only a few species of smooth-barked eucalyptus trees. Branches are grasped between the first 2 and 3 digits of the hand and between the clawless first digit and the remaining digits of the foot. Maturation of a koala takes considerable time. The young koala is carried in the pouch for 6 months and is dependent on its mother for a year. Sexual maturity is not reached for 3 or 4 years. Koalas make sounds including a grunt when feeding and a wail when alarmed.

Phascolarctos cinereus   Koala
O   Qld. Nsw. Vic. Sa.
    Z   W   P.W

                                           P.Z

FAMILY: VOMBATIDAE, Wombats

These marsupials are large and stocky with skull and dentition resembling that of rodents. The teeth are ever growing; the incisors have enamel on the anterior surfaces only. The enamel is much harder and wears more slowly than the dentine on the posterior surface, so the front of the teeth stay sharp for cutting vegetation. There is a wide Diastema (gap) between the incisors and the cheek teeth because there are no canines. The short powerful limbs with broad long claws are used for digging burrows. Wombats are nocturnal and completely herbivorous. The pouch opens to the rear.

Lasiorhinus barnardi   Queensland hairy-nosed wombat (E)
O   Qld.
   Z.   W   P.W
                     P.Z

Lasiorhinus gillespiei   Moonie river wombat
O   S.E. Qld.
   Z.   W.   P.W
                     P.Z

Lasiorhinus latifrons   Southern hairy-nose wombat
O   Sa. To Wa.
   Z.   W.   P.W
                     P.Z

Vombatus ursinus   Common wombat
O   Australia, Tasmania
   Z   W   P.W

                     P.Z

FAMILY:MACROPODIDAE,Kangaroos,Wallaroos,Wallabies,Rat Kangaroos

Macropodids vary in size but all have relatively small heads with large ears. The typical macropodid has very large hind legs with a long powerful tail, structural adaptations for a jumping progression. The hind foot is lengthened and narrow giving the family its name from " macropod " which means large foot. The first syndactylous and digit four is long and strong. Digit five is moderately long. The front legs are short and have 5 digits. Most members of the family graze or browse feeding on many types of plant materials and are ecological equivalents to certain ungulates. The dentition is specialized for their grazing-browsing mode of feeding. There is a high concentration of bacteria in the stomach to aid in the digestion of plant material. The well developed pouch opens forward.

Aepyprymnus rufescens   Rufous rat kangaroo
O   Vic. Nsw. Qld.
   Z   W   P.W
                     P.Z

Bettongia gaimardi   Tasmanian rat kangaroo (E*)
O   Qld. Nsw. Vic.
   Z   W   P.W
                     P.Z.

Bettongia lesueur   Lesueur's rat kangaroo (R)
O   Wa. Sa. Nt. Nsw.
   Z   W   P.W
                     P.Z

Bettongia penicillata
O   S.W.Australia, Nsw.
Z   W

Brush tailed rat kangaroo (E)

P.W
P.Z

Bettongia tropica
O   E.Qld.
Z   W

Northern rat kangaroo (E*)

P.W
P.Z

Caloprymnus campestris
O   Sa. Qld.
Z   W

Desert rat kangaroo (I)

P.W
P.Z

Hypsiprymnodon moschatus
O   N. E. Qld.
Z   W

Musk rat kangaroo

P.W
P.Z

Potorous apicalis
O   Vic. Sa. Tasmania
Z   W

Southern potoroo

P.W
P.Z

Potorous tridactylus
O   Qld. Nsw. Vic.
Z   W

Long nosed potoroo

P.W
P.Z

Dendrolagus bennettianus
O   N.E.Qld.
Z   W

Bennetts tree kangaroo

P.W
P.Z

Dendrolagus dorianus
O   New Guinea, Papua
Z   W

Dorias tree kangaroo (V)

P.W
P.Z

Dendrolagus goodfellowi
O   New Guinea
Z   W

Goodfellows tree kangaroo (V)

P.W
P.Z

Dendrolagus inustus
O   New Guinea
Z   W

Grizzled gray tree kangaroo

P.W
P.Z

Dendrolagus lumholtzi
O   N.E. Qld.
Z   W

Lumholtzs tree kangaroo

P.W
P.Z

Dendrolagus matschiei
O   New Guinea
Z   W

Matschies tree kangaroo

P.W
P.Z

Dorcopsis atrata
O   Goodenough Isl.
Z   W

New Guinea forest wallaby (R)

P.W
P.Z

Dorcopsis hageni   Northern New Guinea wallaby
O   New Guinea
    Z   W   P.W
                                         P.Z

Dorcopsis muelleri   Muellers New Guinea wallaby
O   New Guinea
    Z   W   P.W
                                           P.Z

Dorcopsulus macleayi   New Guinea Mt. wallaby (E)
O   Papua
    Z   W   P.W
                                           P.Z

Dorcopsulus vanheurni   New Guinea Mt. wallaby
O   New Guinea
    Z   W   P.W
                                           P.Z

Lagorchestes conspicillatus   Spectacled hare wallaby
O   Wa. Qld.
    Z   W   P.W
                                           P.Z

Lagorchestes hirsutus   Western hare wallaby (R)
O   Wa.Nt.Sa.
    Z   W   P.W
                                           P.Z

Lagostrophus fasciatus   Banded hare wallaby (R)
O   Wa.
    Z   W   P.W
                                           P.Z

Macropus agilis   River sand wallaby
O   Nt. Qld. Wa. Papua
    Z   W   P.W
                                           P.Z

Macropus antilopinus   Euro
O   Qld. Nt. Wa.
    Z   W   P.W
                                           P.Z

Macropus bernardus   Woodwards wallaroo
O   Arnhem land
    Z   W   P.W
                                           P.Z

Macropus dorsalis   Black striped wallaby
O   Nsw. Qld.
    Z   W   P.W
                                           P.Z

Macropus eugenii   Tammar wallaby
O   S.W.Australia
    Z   W   P.W
                                           P

Macropus fuluginosus   West gray kangaroo (T*)
O   Nsw.Vic. Sa. Wa.
    Z   W   P.W
                                           P.Z

Macropus giganteus   Great gray kangaroo
O   Vic. Qld. Nsw. Tasmania
    Z   W   P.W
                                       P.Z

Macropus Irma   Western scrub wallaby
O   S.W. Australia
    Z   W   P.W
                                       P.Z

Macropus parma   White fronted wallaby (E*)
O   Nsw.
    Z   W   P.W
                                       P.Z

Macropus parryi   Whiptail wallaby
O   Nsw. Qld.
    Z   W   P.W
                                       P.Z

Macropus robustus   Wallaroo
O   Australia
    Z   W   P.W
                                       P.Z

Macropus rufogriseus   Red wallaby
O   Qld. Nsw. Vic. Sa. Tasmania
    Z   W   P.W
                                       P.Z

Megaleia rufus   Red kangaroo (T*)
O   Australia
    Z   W   P.W
                                       P.Z

Onychogalea fraenata   Merin (E)
O   Nsw. Qld.
    Z   W   P.W
                                       P.Z

Onychogalea unguifer   Karrabul
O   N. Australia
    Z   W   P.W
                                       P.Z

Peradorcas concinna   Little rock wallaby
O   Nt.
    Z   W   P.W
                                       P.Z

Petrogale brachyotis   Short eared rock wallaby
O   Nt.
    Z   W   P.W
                                       P.Z

Petrogale godmani   Godmans rock wallaby
O   Cape York peninsul
    Z   W   P.W
                                       P.Z

Petrogale penicillata   Brush tailed rock wallaby
O   Australia
    Z   W   P.W
                                       P.Z

Petrogale purpureicollis   Purple necked rock wallaby
O   Nw. Qld.
    Z   W   P.W
                                                P.Z

Petrogale rothschildi   Rothschilds rock wallaby
O   Wa.
    Z   W   P.W
                                                P.Z

Petrogale xanthopus   Ring tailed rock wallaby (E*)
O   Sa. Nsw.
    Z   W   P.W
                                                P.Z

Setonix brachyurus   Quokka (E*)
O   S.W.Australia
    Z   W   P.W
                                                P.Z

Thylogale billardierii   Red bellied pademelon
O   Sa. Vic Tasmania
    Z   W   P.W
                                                P.Z

Thylogale bruijni   Bruijns pademelon
O   New Guinea, Bismarck Ar.
    Z   W   P.W
                                                P.Z

Thylogale stigmatica   Red legged pademelon
O   Qld. N sw. New Guinea
    Z   W   P.W
                                                P.Z

Thylogale thetis   Red necked pademelon
O   Qld. Nsw.
    Z   W   P.W
                                                P.Z

Wallabia bicolor   Swamp wallaby
O   Qld. Nsw.
    Z   W   P.W
                                                P.Z

TOTAL MARSUPIALIA: O   Z   W   P.W   P.Z

ORDER:INSECTIVORA

An ancient order whose members are close to the generalized primitive mammalian type.

 Insectivores are found throughout much of both hemispheres, but are absent from most of the Australian region, all but the northern most part of South America and the polar regions. Insectivores are generally small animals with small ears and minute eyes which sometimes have no openings at all. The snout is long; the body is covered with short fur and in some instances, spines. They are mainly insectivorous with primitive teeth, though some are carnivorous. Most are nocturnal.

FAMILY: SOLENODONTIDAE, Solenodons

Resemble large, stoutly build shrews.

Solenodon cubanus
O   Cuba

Cuban solenodon (E)

Solenodon paradoxus
O   Haiti

Haitian solenodon (E)

FAMILY: TENRECIDAE, Tenrecs

Found only in Madagascar, having a long, pointed snout, certain species of which are spiny and tailless.

Dasogale fontoynonti
O   E.Madagascar

Tenrec

Echinops telfairi
O   Madagascar

Small Madagascar hedgehog

Hemicentetes nigriceps
O   Madagascar

Streaked tenrec

Hemicentetes semispinosus
O   Madagascar

Streaked tenrec

Setifer setosus
O   Madagascar

Large Madagascar hedgehog

Tenrec ecaudatus
O   Madagascar, Comoro Isl.

Tailless tenrec

Geogale aurita
O   Madagascar

Tenrec

Limnogale mergulus
O   Madagascar

Web footed tenrec

Microgale brevicaudata
O   Madagascar

Tenrec

Microgale cowani
O   Madagascar

Tenrec

Microgale crassipes
O   Madagascar

Tenrec

Microgale decaryi
O   Madagascar, (Fort Dauphin)

Tenrec

Microgale dobsoni
O   Madagascar

Tenrec

Microgale drouhardi O Madagascar	Tenrec
Microgale gracilis O Madagascar	Tenrec
Microgale longicaudata O Madagascar	Long tailed tenrec
Microgale longirostris O Madagascar	Tenrec
Microgale majori O Madagascar	Tenrec
Microgale occidentalis O Madagascar	Tenrec
Microgale parvula O Madagascar , (Maintirano)	Tenrec
Microgale principula O Madagascar	Tenrec
Microgale prolixacaudata O Madagascar	Tenrec
Microgale pusilla O Madagascar	Tenrec
Microgale sorella O E.Madagascar	Tenrec
Microgale taiva O Madagascar, (Tanala)	Tenrec
Microgale talazaci O Madagascar	Tenrec
Microgale thomasi O Madagascar	Tenrec
Oryzorictes hova O Madagascar,(Ankaye)	Rice tenrec
Oryzorictes talpoides O Madagascar	Rice tenrec
Oryzorictes tetradactylus O Madagascar, (Sirabe)	Rice tenrec

FAMILY: POTAMOGALIDAE, Otter shrews

Resembles an otter in appearance

Micropotamogale lamottei   Lesser otter shrew
O   W.Africa

Micropotamogale ruwenzorii   Ruwenzori otter shrew
O   Zaire

Potamogale velox   Giant African otter shrew
O   W. and C. Africa

FAMILY: CHRYSOCHLORIDAE, Golden moles

The fur has an iridescent, red, yellow, bronze, green, often golden luster.

Amblysomus gunningi   Gunnings golden mole
O   South Africa

Amblysomus hottentotus   Hottentot golden mole
O   South Africa

Amblysomus leucorhinus   Congo golden mole
O   C.Africa

Amblysomus obtusirostris   Yellow golden mole
O   S. Africa

Amblysomus sclateri   Sclater's golden mole
O   S. Africa

Amblysomus tropicalis   Tropical golden mole
O   Central Africa

Amblysomus tytonis   Golden mole
O   Somalia

Chrysochloris asiatica   Cape golden mole
O   South Africa

Chrysochloris congicus   Golden mole
O   Central Africa

Chrysochloris fosteri   Foster's golden mole
O   Central Africa

Chrysochloris stuhlmanni   Stuhlmann's golden mole
O   E. Africa

Chrysochloris vermiculus   Golden mole
O   C.Africa

Chrysospalax trevelyani
O   South Africa

Giant golden mole (R)

Chrysospalax villosus
O   South Africa

Rough hair golden mole

Cryptochloris wintoni
O   South Africa (Cape Province)

De Winton's golden mole

Eremitalpa granti
O   South Africa (Cape Province)

Grants desert golden mole

FAMILY : ERINACEIDAE, Hedgehogs

Inhabit the Old World. The backs of most are covered with barbless spines. When threatened, they roll up in a ball bringing the snout and limbs under the body . They live in burrows and appear to have good hearing.

Echinosorex gymnurus
O   Thailand, Sumatra, Borneo

Moon rat

Hylomys suillus
O   S.E.Asia

Lesser gymnure

Neohylomys hainanensis
O   Hainan Isl. (China)

Gymnure

Neotetracus sinensis
O   China, Burma, Indochina

Shrew hedgehog

Podogymnura truei
O   Mindanao, (Philippines)

Mindanao gymnure

Aethechinus angolae
O   Angola

Hedgehog

Atelerix albiventris
O   Africa

Hedgehog

Erinaceus algirus
O   N. Africa ,S.W. Europe

Algerian hedgehog

Erinaceus europaeus
O   Europe, Asia
      Z   W

European hedgehog

P.W

P.Z

Hemiechinus auritus
O   N.Africa, S. Asia

Long eared hedgehog

Hemiechinus megalotis
O   Afghanistan, Russia

Afghan hedgehog

27

Paraechinus aethiopicus   Ethiopian hedgehog
O   Morocco to Arabia to Iraq

Paraechinus hypomelas   Brandt's hedgehog
O   Iran to Arabia

Paraechinus micropus   Indian hedgehog
O   India

FAMILY: TALPIDAE, Moles

Moles have long bodies, a long naked nose and small eyes that often lie beneath the skin. The neck and limbs are very short, the forepaws are strong for digging. Some are semi-aquatic. None are known to hibernate or estivate even though much of their lives are spent underground.

Condylura cristata   Starnosed mole
O   Eastern Canada southward to Georgia
        Z   W   P.W
                                                     P.Z
Desmana moschata   Russian desman (V)
O   Russia

Dymecodon pilirostris   Trues shrew mole
O   Japan, (Hondo Isl.)

Galemys pyrenaicus   Pyrenean desman (R)
O   Spain, Portugal, France

Neurotrichus gibbsii   Shrew mole
O   British Columbia to California

Parascalops breweri   Hairy tailed mole
O   Quebec and Ontario to Ohio

Scalopus aquaticus   Eastern mole
O   Eastern United States

Scalopus inflatus   Tamaulupan mole
O   N. Mexico

Scalopus montanus   Coahiulan mole
O   N.Mexico

Scapanulus oweni   Kansu mole
O   China

Scapanus latimanus   California mole
O   Oregon to Baja California

Scapanus orarius   Pacific mole
O   British Columbia to N.W. California

Scapanus townsendii
O   British Columbia to Baja California

Townsend's mole

Scaptonyx fusicaudus
O   China, Burma

Long tailed mole

Talpa caeca
O   S. Europe, S.W.Asia

Mediterranean mole

Talpa europaea
O   Europe, Asia

Common eurasian mole

Talpa micrura
O   Asia

Eastern mole

Talpa mizura
O   Japan

Japanese mole

Talpa streetorum
O   Iran

Street's mole

Uropsilus soricipes
O   China, Burma

Shrew mole

Urotrichus talpoides
O   Japan

Japanese shrew mole

FAMILY: SORICIDAE, Shrews

Shrews are small, short-legged, mouse-like creatures with long pointed noses. Some live in deserts though most inhabit moist areas of almost all major land areas. They have small eyes and poor vision, but an acute sense of smell and hearing. They are quite active with a high metabolism and can die of fright from loud noises

Anourosorex squamipes
O   S.E.Asia

Szechuan burrowing shrew

Blarina brevicauda
O   S.E. Canada, E.United States

Short tailed shrew

Blarina carolinensis
O   S.E. United States

South short tailed shrew

Blarina telmalestes
O   Virginia, North Carolina

Swamp short tailed shrew

Blarinella quadraticauda
O   China

Short tailed moupin shrew

Chimarrogale phaeura
O   Borneo, Sumatra

Borneo water shrew

Chimarrogale platycephala   Himalayan water shrew
O   Asia

Crocidura aagaardi   Gray shrew
O   S. Asia

Crocidura aequicauda   Gray shrew
O   S.Asia, Sumatra

Crocidura albicauda   White toothed shrew
O   Comoro Isl.

Crocidura allex   White toothed shrew
O   Kenya,Tanzania

Crocidura andamanensis   Andaman shrew
O   Andaman Isl.

Crocidura arethusa   Bauchi musk shrew
O   Nigeria

Crocidura attenuata   Gray shrew
O   Asia

Crocidura baileyi   White toothed shrew
O   Ethiopia, (Simien Mts.)

Crocidura bartelsi   Gray shrew
O   Java

Crocidura batesi   White toothed shrew
O   Gabon, (Como river)

Crocidura beatus   White toothed shrew
O   Philippines

Crocidura beccarii   Gray shrew
O   Sumatra

Crocidura beta   White toothed shrew
O   Kenya, (Fort Hall)

Crocidura bicolor   Tiny musk shrew
O   Botswana to Sudan

Crocidura bloyeti   White toothed shrew
O   E. Africa

Crocidura bolivari   White toothed shrew
O   Africa

Crocidura bottegi   White toothed shrew
O   Ethiopia

Crocidura bovei
O   Zaire

White toothed shrew

Crocidura brevicauda
O   Java

White toothed shrew

Crocidura brunnea
O   Java, Sumatra

Gray shrew

Crocidura buttikoferi
O   Liberia

White toothed shrew

Crocidura butleri
O   Sudan

White toothed shrew

Crocidura caliginea
O   Zaire

White toothed shrew

Crocidura caudata
O   Sicily, Corsica, Balearic Isl.

White toothed shrew

Crocidura cinderella
O   Gambia

White toothed shrew

Crocidura congobelgica
O   Zaire

White toothed shrew

Crocidura crossei
O   Niger

White toothed shrew

Crocidura cyanea
O   Cape to Sudan

Reddish gray musk shrew

Crocidura dolichura
O   Cameroon

White toothed shrew

Crocidura douceti
O   W. Africa

White toothed shrew

Crocidura edwardsiana
O   Philippines

White toothed shrew

Crocidura eisentrauti
O   W. Africa

White toothed shrew

Crocidura elongata
O   Celebes

White toothed shrew

Crocidura ferruginea
O   Sudan

White toothed shrew

Crocidura fischeri
O   Tanzania

White toothed shrew

Crocidura flavescens
O   Cape to Egypt   Giant musk shrew

Crocidura floweri
O   N.E. Africa   Flowers shrew

Crocidura foxi
O   Nigeria   White toothed shrew

Crocidura fuliginosa
O   S.Asia, Borneo   White toothed shrew

Crocidura fulvastra
O   Sudan   White toothed shrew

Crocidura fumosa
O   Kenya to Ethiopia   White toothed shrew

Crocidura fuscosa
O   White Nile   White toothed shrew

Crocidura geoffroyi
O   Mauritania   White toothed shrew

Crocidura giffardi
O   Ivory Coast, Liberia   White toothed shrew

Crocidura glebula
O   N.Nigeria   White toothed shrew

Crocidura goliath
O   Cameroon   Forest shrew

Crocidura gracilipis
O   Tanzania   White toothed shrew

Crocidura grandis
O   Philippines   Shrew

Crocidura grayi
O   Philippines   White toothed shrew

Crocidura halconus
O   Philippines ,(Mindoro)   White toothed shrew

Crocidura hedenborgiana
O   Sudan   White toothed shrew

Crocidura hildegardeae
O   Kenya, Uganda, Ethiopa   White toothed shrew

Crocidura hindei
O   Kenya to Chad   White toothed shrew

Crocidura hirta
O   S.and E. Africa

Zambesi lesser red musk shrew

Crocidura hispida
O   S.Asia

Andaman Isl. spiny shrew

Crocidura horsfieldi
O   S.Asia

Horsfield's shrew

Crocidura indochinensis
O   S. Vietnam

White toothed shrew

Crocidura ingoldbyu
O   W. Africa

White toothed shrew

Crocidura jacksoni
O   Kenya to Zaire

White toothed shrew

Crocidura jouvenetae
O   W. Africa

White toothed shrew

Crocidura langi
O   Zaire

White toothed shrew

Crocidura lasiura
O   Asia

Ussuri large white toothed shrew

Crocidura latona
O   Zaire

White toothed shrew

Crocidura lea
O   Celebes

White toothed shrew

Crocidura lepidura
O   Sumatra

White toothed shrew

Crocidura leucodon
O   Europe, Asia

Bicolor white toothed shrew

Crocidura levicula
O   Celebes

White toothed shrew

Crocidura littoralis
O   Uganda

White toothed shrew

Crocidura ludia
O   Zaire

White toothed shrew

Crocidura luluae
O   Zaire

White toothed shrew

Crocidura luna
O   Zimbabwe, Zaire, Zambia

White toothed shrew

Crocidura lusitania   White toothed shrew
O   Mauritania

Crocidura macarthuri   White toothed shrew
O   Kenya, (Tana river)

Crocidura manni   White toothed shrew
O   Nigeria

Crocidura maporensis   Mapor shrew
O   Mapor Isl. (Malay Peninsula)

Crocidura mariquensis   Black musk shrew
O   Botswana

Crocidura marita   White toothed shrew
O   Sudan

Crocidura martiensseni   White toothed shrew
O   Tanzania, (Marangu)

Crocidura maurisca   White toothed shrew
O   Uganda, Tanzania

Crocidura maxi   White toothed shrew
O   Java

Crocidura melanorhyncha   Shrew
O   Java

Crocidura mindorus   White toothed shrew
O   Philippines

Crocidura minuta   White toothed shrew
O   Java

Crocidura miya   Sri Lanka long tailed shrew
O   Sri Lanka

Crocidura monax   White toothed shrew
O   Tanzania

Crocidura monticola   Sunda shrew
O   Java, Borneo, Timor, Sumba

Crocidura muricauda   White toothed shrew
O   Liberia, (Mt. Coffee)

Crocidura negligens   Koh shrew
O   Koh Isl, (Malay Peninsula)

Crocidura negrina   White toothed shrew
O   Philippines

Crocidura nicobarica
O   Nicobar Isl.

Nicobar shrew

Crocidura nigricans
O   Angola to Sudan

White toothed shrew

Crocidura nigripes
O   Celebes

White toothed shrew

Crocidura nigrofusca
O   Zaire

White toothed shrew

Crocidura nimbae
O   W. Africa

White toothed shrew

Crocidura niobe
O   Uganda

White toothed shrew

Crocidura nyaansae
O   Sudan

White toothed shrew

Crocidura occidentalis
O   Sierre Leone to Ethiopia

Giant musk shrew

Crocidura odorata
O   Gabon

White toothed shrew

Crocidura olivieri
O   N.E. Africa

Egyptyian giant shrew

Crocidura orientalis
O   Java

White toothed shrew

Crocidura oritis
O   Zaire

White toothed shrew

Crocidura palawanensis
O   Philippines, (Palawan)

White toothed shrew

Crocidura paradoxura
O   Sumatra

White toothed shrew

Crocidura parvacauda
O   Celebes, Philippines

White toothed shrew

Crocidura parvipes
O   Kenya

Shrew

Crocidura pasha
O   Sudan

Shrew

Crocidura percivali
O   Kenya

White toothed shrew

Crocidura pergrisea
O   Asia

Pale gray shrew

Crocidura pilosa
O   Cape to Ethiopia

Black musk shrew

Crocidura poensis
O   Gambia to Cameroon

White toothed shrew

Crocidura polia
O   Zaire

White toothed shrew

Crocidura raineyi
O   Kenya

White toothed shrew

Crocidura religiosa
O   N.E.Africa

Egyptian pygmy shrew

Crocidura rhoditis
O   Celebes

White toothed shrew

Crocidura roosevelti
O   Central Africa

White toothed shrew

Crocidura russula
O   Europe, Asia, Africa

Common white toothed shrew

Crocidura zanzibarica
O   Zanzibar

White toothed shrew

Crocidura schweitzeri
O   Liberia

White toothed shrew

Crocidura sericea
O   Sudan

White toothed shrew

Crocidura silacea
O   Africa

White toothed shrew

Crocidura smith
O   Cape to Ethiopia

Desert musk shrew

Crocidura somalica
O   Somali

White toothed shrew

Crocidura suaveolens
O   Europe, Asia, Africa

Lesser white toothed shrew

Crocidura tenuis
O   Timor

White toothed shrew

Crocidura thomensis
O   St. Thome Isl.

White toothed shrew

Crocidura trichura
O   Christmas Isl.

Christmas Isl, shrew

Crocidura ultima
O   Kenya

White toothed shrew

Crocidura velutina
O   Tanzania

White toothed shrew

Crocidura viaria
O   Senegal

White toothed shrew

Crocidura villosa
O   Sumatra

White toothed shrew

Crocidura voi
O   Kenya, (Voi)

White toothed shrew

Crocidura vosmaeri
O   Sumatra, (Banka Isl.)

Banka shrew

Crocidura volcani
O   Cameroon

White toothed shrew

Crocidura weberi
O   Sumatra

White toothed shrew

Crocidura wimmeri
O   W.Africa

White toothed shrew

Crocidura xantippe
O   Kenya, (Voi)

White toothed shrew

Crocidura zaphiri
O   Kenya, Ethiopia

White toothed shrew

Crocidura zimmeri
O   Zaire

White toothed shrew

Cryptotis avius
O   Colombia

Small eared shrew

Cryptotis celatus
O   Mexico, (Oaxaca)

Goodwin's short tailed shrew

Cryptotis endersi
O   Panama

Ender's short tailed shrew

Cryptotis fossor
O   Mexico,(Oaxaca)

Zempoaltepec small eared shrew

Cryptotis goldmani
O   Mexico

Small eared shrew

Cryptotis goodwini   Goodwin's small eared shrew
O   Guatemala

Cryptotis gracilis   Talamancan small eared shrew
O   Costa Rica,Honduras

Cryptotis griseoventris   San Cristobal small eared shrew
O   Mexico, (Chiapas)

Cryptotis guerrerensis   Guerreran small eared shrew
O   Mexico, (Guerrero)

Cryptotis jacksoni   Jackson's small eared shrew
O   Costa Rica

Cryptotis magna   Big small eared shrew
O   Mexico, (Oaxaca)

Cryptotis mera   Mt. Pirri small eared shrew
O   Panama,(Mt. Pirri)

Cryptotis mexicana   Mexican small eared shrew
O   Mexico

Cryptotis micrura   Guatemalan small eared shrew
O   Guatemala

Cryptotis montivagus   Small eared shrew
O   Ecuador

Cryptotis nigrescens   Blackish small eared shrew
O   Honduras to Costa Rica

Cryptotis olivacea   Olivaceous small eared shrew
O   Nicaragua

Cryptotis parva   Least shrew
O   Eastern United States, Mexico to Panama

Cryptotis pergraciles   Slender small eared shrew
O   Mexico

Cryptotis squamipes   Small eared shrew
O   Colombia

Cryptotis surinamensis   Surinam shrew
O   Surinam

Cryptotis tersus   Dark small eared shrew
O   Panama, (Chiriqui)

Cryptotis thomasi   Thomas shrew
O   Ecuador, Venezuela, Colombia

Cryptotis zeteki   Zetek's small eared shrew
O   Panama, (Chiriqui)

Diplomesodon pulchellum   Piebald shrew
O   S. U.S.S.R

Feroculus feroculus   Kelaart's long clawed shrew
O   Sri Lanka

Microsorex hoyi   Pygmy shrew
O   Alaska, Canada, N.E. United States

Microsorex thompsoni   Thompson's pygmy shrew
O   Nova Scotia to Kentucky

Myosorex babaulti   Mouse Shrew
O   E.Africa

Myosorex blarina   Mouse shrew
O   Uganda

Myosorex cafer   Dark footed forest shrew
O   S. Africa

Myosorex eisentrauti   Mouse shrew
O   Fernando Po, Cameroon

Myosorex polli   Mouse shrew
O   Zaire

Myosorex schalleri   Mouse shrew
O   Zaire

Myosorex varius   Mouse shrew
O   South Africa

Nectogale elegans   Szechuan water shrew
O   S. Asia

Neomys anomalus   Water shrew
O   S.Europe, W.Asia

Neomys fodiens   European water shrew
O   Europe, Asia

Notiosorex crawfordi   Crawford's desert shrew
O   S.W. United States, Mexico

Notiosorex gigas   Merriams desert shrew
O   Mexico,(Jalisco)

Paracrocidura shoutedeni   Congo shrew
O   Cameroon, Gabon, Zaire, Congo

Podihik kura   Sri Lanka shrew
O   Sri Lanka

Scutisorex congicus   Congo armoured shrew
O   Zaire

Scutisorex somereni   Uganda armoured shrew
O   Uganda,Rwanda,Zaire

Solisorex pearsoni   Pearson's long clawed shrew
O   Sri Lanka

Sorex alaskanus   Glacier bay water shrew
O   S.E.Alaska

Sorex alpinus   Alpine shrew
O   Europe

Sorex araneus   Common Shrew
O   Europe, Asia

Sorex arcticus   Arctic shrew
O   Alaska to Wisconsin

Sorex bendirii   Pacific water shrew
O   British Columbia to California

Sorex buchariensis   Shrew
O   C. Asia

Sorex caecutiens   Laxmann's shrew
O   Europe, Asia

Sorex cinereus   Masked shrew
O   Siberia, Canada, N.E. Unites States

Sorex cylindricauda   Stripe backed shrew
O   S.E.Asia

Sorex daphaenodon   Shrew
O   Siberia

Sorex dispar   Long tailed shrew
O   Maine to Tennessee

Sorex fumeus   Smoky shrew
O   S.E.Canada, N.E. United States

Sorex gaspensis   Gaspe shrew
O   Quebec, (Gaspe peninsula)

Sorex hawkeri   Asiatic pygmy shrew
O   Asia

Sorex hydrodromus   Unalaska shrew
O   Alaska, (St. Paul Isl.)

Sorex jacksoni   St. Lawrence Isl. shrew
O   St. Lawrence Isl.

Sorex juncensis   Tulu shrew
O   Baja California,(San Quintin)

Sorex longirostris   Southeastern shrew
O   S.E. United States

Sorex lyelli   Mount Lyell shrew
O   N.E. California

Sorex macrodon   Large toothed shrew
O   Mexico, (Veracruz)

Sorex merriami   Merriam shrew
O   W. United States

Sorex milleri   Carmen Mt. shrew
O   Mexico, (Coahuila)

Sorex minutus   Lesser shrew
O   Europe, Asia

Sorex nanus   Dwarf shrew
O   Montana to Arizona and New Mexico

Sorex obscurus   Dusky shrew
O   Alaska, Canada, United States

Sorex oeropolus   Mexican long tailed shrew
O   Mexico

Sorex ornatus   Ornate shrew
O   California, Baja California

Sorex pacificus   Pacific shrew
O   W.Oregon, N.W. California

Sorex palustris   Water shrew
O   Alaska to New Mexico and east to Nova Scotia

Sorex preblei   Preble's shrew
O   E. Oregon

Sorex pribilofensis   Pribilof shrew
O   Alaska,(St.Paul Isl.)

Sorex saussurei   Saussure's shrew
O   Mexico, Guatemala

Sorex sclateri   Sclater's shrew
O   Mexico, (Chiapas)

Sorex sinousus   Suisun shrew
O   California, (Grizzly Isl.)

Sorex stizodon   San Cristobal shrew
O   Mexico, (Chiapas)

Sorex tenellus   Inyo shrew
O   Nevada, California

Sorex trigonirostris   Ashland shrew
O   Oregon

Sorex trowbridgeii   Trowbridge's shrew
O   British Columbia to California

Sorex vagrans   Vagrant shrew
O   Alaska to Mexico

Sorex veraepacis   Verapaz shrew
O   Mexico to Guatemala

Soriculus caudatus   Hodgsins brown toothed shrew
O   Asia

Soriculus gruberi   Grubers shrew
O   Nepal

Soriculus hypsibius   De Winton's shrew
O   S.Asia

Soriculus leucops   Indian long tailed shrew
O   S.Asia

Soriculus lowei   Lowe's shrew
O   China

Soriculus nigrescens   Sikkim large clawed shrew
O   S.Asia

Soriculus parva   Shrew
O   China, (Szechwan)

Soriculus salenskii   Salenski's shrew
O   S.Asia

Suncus ater   Black shrew
O   Borneo,(Kinabalu)

Suncus dayi   Days shrew
O   S. India

Suncus etruscus O S.Europe, Asia	Savis pygmy shrew
Suncus infinitesimus O C. and S. Africa	Shrew
Suncus leucura O E.Africa	Shrew
Suncus lixus O Zaire, Kenya to South Africa	Greater dwarf shrew
Suncus luzoniensis O Philippines	Shrew
Suncus madagascariensis O Madagascar	Madagascar shrew
Suncus murinus O Africa, Asia minor, S.Asia	House shrew
Suncus occultidens O Philippines	Shrew
Suncus ollula O W. Africa	Shrew
Suncus palawanensis O Philippines,(Palawan)	Shrew
Suncus remyi O Gabon	Shrew
Suncus ruandae O C. Africa	Shrew
Suncus stoliczkanus O S.Asia	Anderson's shrew
Suncus varilla O Cape Province to Malawi	Shrew
Surdisorex norae O Kenya,(Aberdare)	Mt. Shrew
Surdisorex polulus O Kenya, (Mt. Kenya)	Mt. Shrew
Sylvisorex granti O Cameroon to Tanzania	Grant' shrew
Sylvisorex johnstoni O W. Africa	Johnston's shrew

Sylvisorex lunaris	Shrew
O Uganda,Zaire	

Sylvisorex megalura	Shrew
O Liberia	

Sylvisorex morio	Shrew
O Nigeria	

Sylvisorex oriundus	Shrew
O Congo	

Sylvisorex preussi	Shrew
O Cameroon, Nigeria	

Sylvisorex suncoides	Shrew
O C. Africa	

FAMILY: MACROSCELIDIDAE,Elephant shrews

The elephant shrews are found exclusively in Africa. The name refers to the resemblance between the trunk of an elephant and the long, flexible snouts of these shrews. Found mainly in rocky areas, dikes or crevices and prefer well drained soils. They often live in burrows started by small rodents or in the crevices of termite hills. Unlike most insectivores, the elephant shrews are specialized for jumping. The long narrow snout is extremely sensitive and mobile with the nostril at the tip. It is used in searching for food and in pushing and lifting debris to search for ants.Scent glands under the tail and on the chest are used for marking territories. Extremely alert and agile, the elephant shrew will dash swiftly for cover when alarmed. Tapping of hind legs and squealing are common alarm signals. Lifespan in the wild may only be about 18 months, but in captivity is probably longer.

Elephantulus brachyrhynchus	Short snouted elephant shrew
O Botswana	

Elephantulus edwardi	Elephant shrew
O Cape Province	

Elephantulus fuscipes	Elephant shrew
O Uganda,Zaire	

Elephantulus intufi	Bushveid elephant shrew
O S. Africa	

Elephantulus myurus	Rock elephant shrew
O Botswana	

Elephantulus revoilii	Elephant shrew
O Somalia	

Elephantulus rozeti	North African elephant shrew
O Morocco, Algeria	

Elephantulus rufescens O Mozambique to Kenya	Elephant shrew
Elephantulus rupestris O S. Africa	Rock elephant
Macroscelides proboscideus O S.Africa	Short eared elephant shrew
Petrodromus tetradactylus O E. and C.Africa	Four toed elephant shrew
Rhynchocyon chrysopygus O Kenya	Elephant shrew
Rhynchocyon cirnei O E.Africa	Checkered elephant shrew
Rhynchocyon peterdi O Tanzania, Kenya, Zanzibar	Elephant shrew

FAMILY: TUPAIIDAE, Tree shrews

Tree shrews are small, squirrel-like animals having slender bodies, long, usually bushy tails and long noses. They live in the tropical rain forests of Southeast Asia and surrounding islands. Claws are present on all digits and hands are convergent. They are good climbers and swift runners. Vision is fair, (eyes toward the side of the head) but senses of smell and hearing are keen. Diet consists of insects, grubs and fruit. Until recently, tree shrews have been classified in the Order Primates and many reference books will list them in this way.

Anathana ellioti O India	Madras tree shrew
Dendrogale melanura O Borneo	Smooth tailed tree shrew
Dendrogale murina O S.E.Asia	Smooth tailed tree shrew
Ptilocercus lowii O Malaysia, Borneo, Sumatra	Pen tailed tree shrew
Tupaia dorsalis O Borneo	Striped tree shrew
Tupaia glis O China, S.E.Asia	Common tree shrew
Tupaia gracilis O Borneo	Slender tree shrew

45

Tupaia javanica
O   Java, Sumatra

Small tree shrew

Tupaia minor
O   S.E.Asia

Lesser tree shrew

Tupaia montana
O   Borneo

Mountain tree shrew

Tupaia nicobarica
O   Nicobar Isl.

Nicobar tree shrew

Tupaia palawanensis
O   Philippines

Palawan tree shrew

Tupaia picta
O   Borneo

Painted tree shrew

Tupaia splendidula
O   Borneo

Red tailed tree shrew

Tupaia tana
O   Sumatra, Borneo

Large tree shrew

Urogale everetti
O   Philippines, (Mindanao)

Philippine tree shrew

TOTAL: INSECTIVORA: O	Z	W	P.W	P.Z

46

ORDER:DERMOPTERA

Called "flying lemurs" ,the animals of this order neither fly nor are they lemurs. They resemble lemurs in appearance and glide from branch to branch by means of a large membrane attached to the neck and sides of the body, extending "cape fashion" along the limbs to the tips of the fingers, toes and tail. Dermopterans are skillful climbers well adapted for arboreal living but are nearly helpless on the ground. Active at night, they feed on vegetable matter. Dermopterans live in tropical forests of South Burma, South Indochina, Malaysia, Sumatra, Bornea, Java, and some of the Philippine islands.

FAMILY: CYNOCEPHALIDAE, Flying Lemurs

Cynocephalus volans   Flying lemurs
O   Philippines

Cynocephalus variegatus   Flying lemurs
O   S.E. Asia

ORDER:CHIROPTERA

Bats are the only mammals capable of true flight. They are a highly successful group; among mammals, only rodents have more species. The distribution of bats is nearly cosmopolitan. They are absent only from the arctic and polar regions and some isolated oceanic islands, however, most bats inhabit subtropical and tropical areas.Bats are well adapted for flight. The second through fifth digit of the forelimb are elongated to support the wing membrane which is an extension of skin of back and belly attached to the arms, sides of the , legs, and tail. The membranes are elastic, thin and consist only of skin; there is no flesh in between. The first digit is short and has a sharp hooked claw (some 2 claws) used to cling to surfaces.Bats are divided into two suborders, Magachiroptera contains one family, Pteropodidae, fruit and nectar eating bats of the Old World. Some of these are called "flying foxes" because they have fox-like faces and are quite large. All other families belong to the suborder Microchiroptera. There are two important functional differences between the suborders. Megachiropterans are not known to hibernate. They maintain the body temperature within narrow limits by physiological and behavioral means. Many Microchiropterans are heterothermic. Some hibernate for long periods. Microchiropterans find their way about by echolocation. This is a phenomenon in which the bat, as it flies, emits high pitched vocal sounds through the mouth or nose usually above the limit of human hearing. The sounds are reflected back in the form of echoes. The echoes enable the bat to avoid hitting objects and to find flying insects. The refinement of the echolocation system is varied, being most developed in insectivorous bats. Megachiropterans rely on vision and are helpless in total darkness. Bats shelter in caves, crevices, tree cavities, and buildings. They normally rest hanging head down and take flight easily from that position, literally "swimming" through the air using the legs and arms in unison. Usually only one young is born per year. This slow reproductive rate is offset by the relatively long lifespan,as long as 17 years in some forms.

For easier study, bats are sometimes grouped by feeding habits which are very diverse. Bats have radiated to fill several niches, some of which are occupied by birds during the day.

1.   Insectivorous - Small in size, includes the majority of bats. Feed on the wing, sometimes also eat fruit.
2.   Fructivorous - Feed almost entirely on fruit, possibly a little vegetation. Live in tropics following ripening fruit. Wing span ranges from ten inches to the five foot span of the "flying fox".
3.   Flower feeders - Feed on pollen and nectar. Small long pointed heads, long tongues with brush-like tips. Inhabit tropics or semi-tropics.
4.   True vampires - Live on blood lapped from a small incision made in the skin of a sleeping animal, Can transmit diseases. Only three species in the group.
5.   Carnivorous - Prey on other small mammals, birds, lizards, frogs. Have widely varied diet. Moderate size.
6.   Fish-eating - Catch fish near water's surface using large powerful feet with hooked claws

SUBORDER: MEGACHIROPTERA
FAMILY: PTEROPODIDAE, Flying foxes, Old world fruit bats

Acerodon celebensis   Fruit bat
O   Celebes, Sula Isl.

Acerodon humilis   Fruit bat
O   Talaut Isl.

Acerodon jubatus   Fruit bat
O   Philippines

Acerodon lucifer   Fruit bat
O   Philippines, (Panay Isl.)

Acerodon mackloti   Fruit bat
O   Timor, Flores

Aethalop alecto   Gray fruit bat
O   Sumatra, Malaya, Borneo

Balionycteris maculata   Spotted winged fruit bat
O   S.Asia, Borneo

Boneia bidens   Flying fox
O   Celebes

Casinycteris argynnis   Short-Palate fruit bat
O   Cameroon to Zaire

Chironax melanocephalus   Black capped fruit bat
O   S.Asia, Sumatra, Java

Cynopterus archipelagus   Fruit bat
O   Philippines

Cynopterus brachyotis   Malasian fruit bat
O   Sri Lanka to Philippines

Cynopterus horsfieldi   Horsfields fruit bat
O   S.E.Asia

Cynopterus sphinx   Short nosed fruit bat
O   Pakistan to Java, Bali

Dobsonia crenuluta   Bare backed fruit bat
O   Halmahera Isl.

Dobsonia exoleta   Bare backed fruit bat
O   Celebes

Dobsonia inermis   Bare backed fruit bat
O   Solomon Isl.

Dobsonia minor   Bare backed fruit bat
O   New Guinea

Dobsonia moluccensis   Bare backed fruit bat
O   Australia, New guinea

Dobsonia peroni   Bare backed fruit bat
O   Timor, Bali, Sumatra

Dobsonia praedatrix   Bare backed fruit bat
O   New Britain

Dobsonia remota   Bare backed fruit bat
O   Trobriand Isl.

Dobsonia viridis   Bare backed fruit bat
O   Kei, Banda, Amb

Dyacopterus spadiceus   Dayak fruit bat
O   Malay peninsula, Sumatra, Borneo

Eidolon helvum   Straw colored fruit bat
O   Africa, Madagascar

Epomophorus angolensis   Angolan epauletted bat
O   Angola

Epomophorus anurus   Eastern epauletted bat
O   Ethiopia to Niger

Epomophorus crypturus   Peters epauletted bat
O   Zaire to S.Africa

Epomophorus gambianus   Gambian epauletted bat
O   Senegal to Zambia

Epomophorus labiatus   Little epauletted fruit bat
O   Ethiopia to Zambia

Epomophorus pousarguesi   Pousargues epauletted bat
O   Upper Chari river

Epomophorus reii   Garua epauletted bat
O   Cameroon

Epomophorus wahlbergi   Wahlberg epauletted bat
O   Cameroon and Somalia, to S.Africa

Epomops buttikoferi   Buttikofers fruit bat
O   Guinea to Nigeria

Epomops dobsoni   Dobsons fruit bat
O   Angola, Zambia, Zaire, Tanzania

Epomops franqueti   Franquets fruit bat
O   Ivory Coast to Zambia

Haplonycteris fischeri   Fruit bat
O   Philippines, (Mindoro)

Hypsignathus monstrosus   Hammer headed fruit bat
O   Gambia to Kenya

Megaerops ecaudatus   Tailless fruit bat
O   S.Asia to Borneo

Megaerops wetmorei   Fruit bat
O   Philippines, (Cotabato)

Micropteropus grandis   Sanborns epauletted fruit bat
O   Angola

Micropteropus intermedius   Haymans epauletted bat
O   Angola, Zaire

Micropteropus pusillus   Dwarf epauletted fruit bat
O   Senegal to Tanzania

Myonycteris brachycephala   Sao Tome collared fruit bat
O   Sao Tome Isl.

Myonycteris torquata   Little collared fruit bat
O   Sierra Leone to Zambia

Nanonycteris veldkampi   Veldkamps dwarf fruit bat
O   Guinea to Cameroon

Neoperyx frosti   Fruit bat
O   Celebes

Penthetor lucasi   Dusky fruit bat
O   S.Asia to Borneo

Plerotes anchietai   Anchietas fruit bat
O   Angola, Zambia

Ptenochirus jagori   Fruit bat
O   Borneo, Philippines

Pteralopex atrata   Fruit bat
O   Solomon Isl.

Pteropus admiralitatum   Flying fox
O   Admiralty, Solomon Isl.

Pteropus aldabrensis   Aldabra flying fox
O   Aldabra Isl.

Pteropus alecto   Flying fox
O   Australia, Celebes

Pteropus argentatus   Flying fox
O   Celebes

Pteropus arquatus   Flying fox
O   Celebes

Pteropus balutus   Flying fox
O   Philippines,Balut Isl.

Pteropus caniceps   Flying fox
O   Halmahera Isl, Celebes

Pteropus chrysoproctus   Flying fox
O   Amboina, Buru, Ceram

Pteropus cognatus   Flying fox
O   Solomon Isl.

Pteropus comorensis   Comoro mantled flying fox
O   Comoro Isl.

Pteropus conspicillatus   Spectacled flying fox
O   Australia, New Guinea

Pteropus dasymullus   Liukiu Isl, flying fox
O   Liukiu Isl. , Taiwan

Pteropus dobsoni   Flying fox
O   Celebes

Pteropus giganteus   Indian flying fox
O   Sri Lanka, India to Burma

Pteropus gilliardi   Flying fox
O   New Britain

Pteropus griseus   Flying fox
O   Celebes, Timor

Pteropus hypomelanus   Island flying fox
O   S.E.Asia, Philippines

Pteropus leucopterus   Flying fox
O   Philippines, (Luzon)

Pteropus livingstonei   Comoro black flying fox
O   Johanna Isl.

Pteropus lombocensis   Flying fox
O   Lombok, Flores

Pteropus lylei   Large flying fox
O   Thailand, South Vietnam

Pteropus macrotis   Flying fox
O   Aru Isl., New Guinea

Pteropus mahaganus   Flying fox
O   Solomon Isl.

Pteropus mariannus   Flying fox (V)
O   Mariana Isl, Liukiu Isl.

Pteropus mearnsi   Flying fox
O   Philippines,(Mindanao)

Pteropus melanopogon   Flying fox
O   Aru, Kei, Banda, Ceram Isl.

Pteropus melanotus   Nicobar flying fox
O   Nicobar, Andaman Isl.

Pteropus mimus   Flying fox
O   Philipines, Celebes

Pteropus neohibornicus   Flying fox
O   Bismark Arch, Admiralty Isl.

Pteropus niger   Greater Mascarene flying fox (R)
O   Mascarenes

Pteropus ocularis   Flying fox
O   Cerum, buru

Pteropus personatus   Flying fox
O   N.Celebes

Pteropus pohlei   Flying fox
O   New Guinea

Pteropus poliocephalus   Gray headed flying fox
O   Australia

Pteropus pumilus   Flying fox
O   Philippines, (Palmas Isl.)

Pteropus rayneri   Flying fox
O   Solomon Isl.

Pteropus rodricensis   Rodriguez flying fox (E)
O   Rodriguez Isl.

Pteropus rufus
O   Madagascar

Madagascar flying fox

Pteropus scapulatus
O   Australia

Red flying fox

Pteropus seychellensis
O   Seychelles

Seychelles flying fox

Pteropus speciosus
O   Philippines, (Malanipa Isl)

Flying fox

Pteropus subniger
O   Mascarenes

Lesser Mascarene flying fox

Pteropus tablasi
O   Philippines

Flying fox

Pteropus temmincki
O   Amboina, Ceram, Timor, Bu

Flying fox

Pteropus tonganus
O   Solomon Isl., New Guinea

Flying fox

Pteropus vampyrus
O   Burma to Philippines

Malayan flying fox

Pteropus voeltzkowi
O   Solomon Isl.

Pemba flying fox

Pteropus woodfordi
O   Solomon Isl.

Flying fox

Rousettus aegyptiacus
O   Cape prov. to Egypt

Egyptian fruit bat

Rousettus amplexicaudatus
O   S.E.Asia, Celebes, New Guinea

Geogroys rousette

Rousettus angolensis
O   Guinea to Zambia

Bocages fruit bat

Rousettus celebensis
O   Celebes,(Sanghir Isl.)

Dog bat

Rousettus lanosus
O   Uganda, Tanzania

Ruwenzori long hair rousette

Rousettus leschenaulti
O   India to Viet Nam, S. China, Java

Fruit bat

Rousettus seminudus
O   Sri Lanka

Fruit bat

Rousettus stresemanni   Dog bat
O New Guinea

Scotonycteris ophiodon   Pohles fruit bat
O Liberia to Cameroon

Scotonycteris zenkeri   Zenkers fruit bat
O Liberia to Zaire

Sphaerias blanfordi   Blanfords fruit bat
O Burma, Thailand

Styloctenium wallacei   Fruit bat
O Celebes

Thoopterus nigrescens   Short nosed fruit bat
O Celebes, (Hal Isl.)

Eonycteris major   Cave fruit bat
O Borneo

Eonycteris robusta   Fruit bat
O Philippines, (Luzon)

Eonycteris rosenbergi   Fruit bat
O N.Celebes

Eonycteris spelaea   Long tailed fruit bat
O S.E.Asia to Borneo

Macroglossus fructivorus   Long tongued fruit bat
O Philippines

Macroglossus lagochilus   Northern blossom bat
O Australia to Malaysia

Macroglossus minimus   Long tongued fruit bat
O Thailand, Burma to Sumatra

Magaloglossus woermanni   African long tongued fruit bat
O Guinea to Uganda

Melonycteris melanops   Long tongued bat
O Papua, Bismark Arch.

Melonycteris woodfordi   Long tongued bat
O Solomon Isl

Notopteris macdonaldi   Long tailed fruit bat
O W. Polynesia

Syconycteris australis   Blossom bat
O Australia

Syconycteris crassa   Blossom bat
O   New Guinea, Trobriand Isl.

Syconycteris naias   Blossom bat
O   Trobriand Isl,

Nyctimene aello   Tube nosed bat
O   New Guinea

Nyctimene albiventer   Papuan tube nosed bat
O   Australia, New Guinea, Solomon Isl

Nyctimene cephalotes   Tube nosed bat
O   Celebes, Timor, Cerma, New Guinea

Nyctimene cyclotis   Tube nosed bat
O   S.New Guinea

Nyctimene major   Tube nosed bat
O   New Guinea, Solomon Isl.

Nyctimene minutus   Tube nosed bat
O   Celebes,(Amboina Isl)

Nyctimene robinsoni   Queensland tube nosed bat
O   Qld.

Paranyctimene raptor   Lesser tube nosed bat
O   New Guinea

Harpyionycteris whiteheadi   Harpy fruit bat
O   Celebes, Philippines

SUBORDER: MICROCHIROPTERA

FAMILY: RHINOPOMATIDAE, Mouse tailed bats

Rhinopoma hardwickei   Lesser mouse tailed bat
O   Morocco,Kenya to India

Rhinopoma microphyllum   Large mouse tailed bat
O   Senegal,Nigeria to India

FAMILY:EMBALLONURIDAE, Sheath tailed bats

Balantiopteryx infusca   Sac winged bat
O   Ecuador

Balantiopteryx io   Sac winged bat
O   S.Mexico, Guatemala

Balantiopteryx plicata
O   N.Mexico to Costa Rica

Peters bat

Centromyucteris maximiliani
O   Yucatan to N. South America

Thomas bat

Coleura afra
O   Africa

Mozambique sheath tailed bat

Coleura seychellensis
O   Seychelles

Seychelles sheath tailed bat

Cormura brevirostris
O   Nicaragua to Brazil

Wagners sac winged bat

Emballonura alecto
O   Celebes, Philippines

Sheath tailed bat

Emballonura atrata
O   Madagascar

Peters sheath tailed bat

Emballonura beccarii
O   New Guinea, Papua

Sheath tailed bat

Emballonura furax
O   New Guinea

Sheath tailed bat

Emballonura monticola
O   Burma to Borneo

Lesser sheath tailed bat

Emballonura nigrescens
O   Solomons, Amboina

Sheath tailed bat

Emballonura papuana
O   Celebes, New Guinea

Sheath tailed bat

Emballonura raffrayana
O   New Guinea, Solomons

Sheath tailed bat

Emballonura rivalis
O   Borneo

Sheath tailed bat

Peroperyx kappeleri
O   Guatemala to Peru

Sheath tailed bat

Peropteryx leucopterus
O   Venezuela to Peru

Dog like bat

Peropteryx macrotis
O   Yucatan to Sao Paulo

Lesser dog like bat

Rhynchonycteris naso
O   S.Mexico to Peru

Proboscis bat

Saccopteryx bilineata | White lined bat
O   S.Mexico to Peru

Saccopteryx canescens | White lined bat
O   Colombia to Paraguay

Saccopteryx gymnura | White lined bat
O   Brazil

Saccopteryx leptura | Lesser white lined bat
O   Chiapas to Peru

Saccopteryx pumilis | White lined bat
O   Venezuela, Guianas

Taphozous australis | Tomb bat
O   Australia

Taphozous flaviventris | Yellow bellied sheath tailed bat
O   Australia

Taphozous georgianus | Common sheath tailed bat
O   Australia

Taphozous hamiltoni | Hamiltons tomb bat
O   Sudan

Taphozous hildegardeae | Hildegardes tomb bat
O   Kenya

Taphozous kachhensis | Tomb bat
O   India, Iraq, Burma, Malaysia

Taphozous longimanus | Long winged tomb bat
O   Sri Lanka to Borneo

Taphozous mauritianus | Mauritian tomb bat
O   Madagascar, Africa

Taphozous melanopogon | Black bearded tomb bat
O   India to Borneo

Taphozous mixtus | New Guinea sheath tailed bat
O   Australia

Taphozous nudicluniatus | Naked rumped sheath tailed bat
O   Australia

Tophozous nidiventris | Naked rumped tomb bat
O   India to Egypt to Tanzania

Taphozous peli | Pels pouched bat
O   Liberia to Kenya

Taphozous perforatus   Egyptian tomb bat
O   Egypt to Rhodes, India

Taphozous philippensis   Sheath tailed bat
O   Philippines, (Mindanao)

Taphozous saccolaimus   Pouch bearing bat
O   Sri Lanka to Borneo

Taphozous theobaldi   Tomb bat
O   India to S.E.Asia

Cyttarops alecto   Ghost bat
O   Nicaragua to Guianas

Depanycteris isabella   Ghost bat
O   Amazonas,Brazil

Diclidurus albus   Ghost bat
O   Mexico to Brazil

Diclidurus ingens   Ghost bat
O   Colombia

Diclidurus scutatus   Ghost bat
O   Peru,Venezuela

FAMILY: NOCTILIONDAE,Bulldog bats

Noctilio labialis   Southern bulldog bat
O   Nicaragua to Argentina

Noctilio leporinus   Bulldog bat
O   Mexico to Argentina

FAMILY: NYCTERIDAE, Slit faced bats

Nycteris aethiopoica   Ethiopian slit faced bat
O   E.Africa

Nycteris arge   Bates slit faced bat
O   Sierra Leone to Kenya

Nycteris gambiensis   Gambian slit faced bat
O   Senegal to Sierra Leone

Nycteris grandis   Large slit faced bat
O   Guinea to Zambia

Nycteris hispida   Hairy slit faced bat
O   Africa

Nycteris javanicus
O   Burma to Borneo

Hollow faced bat

Nycteris macrotis
O   Gambia to Zambia

Large eared slit faced bat

Nycteris major
O   Cameroon, Zaire

Ja slit faced bat

Nycteris nana
O   Ghana to Kenya

Dwarf slit faced bat

Nycteris parissi
O   Somalia, Cameroon

Paris slit faced bat

Nycteris thebaica
O   Africa, Arabia

Egyptian slit faced bat

Nycteris woodi
O   Zambia, Zimbabwe

Woods slit faced bat

FAMILY: MEGADERMATIDAE, False Vampire bats

Cardioderma cor
O   E. Africa

Heart nosed big eared bat

Lavia frons
O   Gambia to E.Africa

Yellow winged bat

Macroderma gigas
O   Australia

Australian false vampire (V)

Megaderma lyra
O   India, Malaysia

Indian false vampire

Megaderma spasma
O   India to Philippines

False vampire

FAMILY:RHINOLOPHIDAE, Horseshoe and Leaf Nosed bats

Anthops ornatus
O   Solomon Isl.

Flower faced bat

Asellia patrizii
O   Ethiopoia

Leaf nosed bat

Asellia tridens
O   Gambia to Iraq

Trident leaf nosed bat

Rhinolophus acuminatus
O   Java, Lombok, Borneo

Horseshoe bat

Rhinolophus adami
O   Congo

Horseshoe bat

Rhinolophus affinis
O   Burma to Java

Intermediate horseshoe bat

Rhinolophus alcyone
O   W. Africa

Halcyon horseshoe bat

Rhinolophus anderseni
O   Philippines

Horseshoe bat

Rhinolophus arcuatus
O   Philippines, Borneo

Horseshoe bat

Rhinolophus blasii
O   Africa, Europe

Peak saddle horseshoe bat

Rhinolophus blythi
O   China

Leaf nosed bat

Rhinolophus borneensis
O   Borneo, Cambodia

Horseshoe bat

Rhinolophus canuti
O   Java

Horseshoe bat

Rhinolophus capensis
O   S. Africa

Cape horseshoe bat

Rhinolophus celebensis
O   Celebes

Celebes horseshoe bat

Rhinolophus clivosus
O   Africa

Geoffroy's horseshoe bat

Rhinolophus coelophyllus
O   Burma, Malaya

Croslet horseshoe bat

Rhinolophus cornutus
O   Japan

Japan horseshoe bat

Rhinolophus creaghi
O   Borneo

Horseshoe bat

Rhinolophus darlingi
O   S.Africa to Tanzania

Darlings horseshoe bat

Rhinolophus denti
O   S.W.Africa

Dents horseshoe bat

Rhinolophus episcopus
O   China

Leaf nosed bat

Rhinolophus euryale   Mediterranean horseshoe bat
O   N.Africa, Europe

Rhinolophus euryotis   Horseshoe bat
O   Amboina, Ceram, New Guinea

Rhinolophus ferrumequinum   Greater horseshoe bat
O   Palearctic

Rhinolophus fumigatus   Ruppells horseshoe bat
O   Africa

Rhinolophus hidebrandti   Hildebrandts horseshoe bat
O   Africa

Rhinolophus hipposideros   Lesser horseshoe bat
O   Africa

Rhinolophus hirsutus   Horseshoe bat
O   Philippines,(Guimaras Isl)

Rhinolophus importunis   Horseshoe bat
O   Java

Rhinolophus inops   Horseshoe bat
O   Philippines

Rhinolophus javanicus   Horseshoe bat
O   S.Java

Rhinolophus keyensis   Horseshoe bat
O   Kei, Halmahera Isl.

Rhinolophus landeri   Landers horseshoe bat
O   Gambia to Congo

Rhinolophus lanosus   Landers horseshoe bat
O   China

Rhinolophus lepidus   Horseshoe bat
O   China, Burma

Rhinolophus luctus   Woolly horseshoe bat
O   India to Borneo

Rhinolophus maclaudi   Horseshoe bat
O   Guinea

Rhinolophus macrotis   Big eared horseshoe bat
O   Nepal to Philippines

Rhinolophus madurensis   Horseshoe bat
O   Madura Isl.

Rhinolophus malayanus
O   Laos to Malaysia

North Malayan horseshoe bat

Rhinolophus megaphyllus
O   Australia, Papua

Eastern horseshoe bat

Rhinolophus mehelyi
O   N.Africa, S.Europe

Mehelys horseshoe bat

Rhinolophus mindanao
O   Philippines

Horseshoe bat

Rhinolophus minutillus
O   Thailand to Malaysia

Least horseshoe bat

Rhinolophus monoceros
O   Taiwan

Horseshoe bat

Rhinolophus nereis
O   Anamba, Natuna Isl.

Horseshoe bat

Rhinolophus pearsoni
O   India, China, Vietnam

Horseshoe bat

Rhinolophus philippinensis
O   Australia to Philippines

Large eared horseshoe bat

Rhinolophus pusillus
O   Java

Horseshoe bat

Rhinolophus refulgens
O   Sumatra, Malaysia

Glossy horseshoe bat

Rhinolophus rex
O   China

Horseshoe bat

Rhinolophus robinsoni
O   Malaysia, Thailand

Peninsular horseshoe bat

Rhinolophus rouxi
O   Sri Lanka, India, S.China

Horseshoe bat

Rhinolophus ruwenzorii
O   E.Congo,Zaire

Ruwenzori horseshoe bat

Rhinolophus sedulus
O   Malaysia, Borneo

Lesser woolly horseshoe bat

Rhinolophus shameli
O   Thailand, Cambodia

Horseshoe bat

Rhinolophus simplex
O   Lombox, Sumbawa

Horseshoe bat

Rhinolophus simulator
O   S.Africa to Sudan

Bushveld horseshoe bat

Rhinolophus stheno
O   Malaysia to Java

Lesser brown horseshoe bat

Rhinolophus subbadius
O   Nepal, India,Viet Nam

Horseshoe bat

Rhinolophus subrufus
O   Philippines, (Luzon)

Horseshoe bat

Rhinolophus swinnyi
O   S. Africa

Swinny's horseshoe bat

Rhinolophus thomasl
O   Burma, Yunnan, Viet Nam

Thomas horseshoe bat

Rhinolophus toxopeusi
O   Buru

Horseshoe bat

Rhinolophus trifoliatus
O   Burma to Java

Trefoil horseshoe bat

Rhinolophus virgo
O   Philippines, (Luzon)

Horseshoe bat

Rhinomegalophus paradoxolophus
O   S.Asia

Horseshoe bat

Aselliscus stoliczkanus
O   Burma to China

Trident horseshoe bat

Aselliscus tricuspidsatus
O   New Guinea

Tates trident bat

Cloeotis percivali
O   E. and S. Africa

Short eared trident bat

Coelops frithii
O   India to Java

Tailless horseshoe bat

Coelops hirsutus
O   Philippines

Horseshoe bat

Coelops robinsoni
O   Thailand

Malayan tailless horseshoe bat

Hipposideros abae
O   W. Africa to Uganda

Aba leaf nosed bat

Hipposideros armiger
O   Burma to S. China

Leaf nosed bat

Hipposideros ater
O   Australia

Dusky horseshoe bat

Hipposideros beatus
O   Sierre Leone to Sudan

Dwarf leaf nosed bat

Hipposideros bicolor
O   India to Philippines

Bicolor leaf nosed bat

Hipposideros caffer
O   Africa

Sundevall's African leaf nosed bat

Hipposideros calcaratus
O   Solomon, New Guinea

Leaf nosed bat

Hipposideros camerunensis
O   Mt. Cameroon

Greater cyclops bat

Hipposideros cervinus
O   Celebes, New Guinea

Leaf nosed bat

Hipposideros cineraceus
O   India to Malaysia

Leaf nosed bat

Hipposideros commersoni
O   Africa

Commerson's leaf nosed bat

Hipposideros coxi
O   Borneo

Roundleaf horseshoe bat

Hipposideros cupidus
O   New Guinea, Bismark, Solomon

Leaf nosed bat

Hipposideros curtus
O   Cameroon

Short tailed leaf nosed bat

Hipposideros crumeniferus
O   Timor

Leaf nosed bat

Hipposideros cyclops
O   Guinea to Uganda

Cyclops bat

Hipposideros diadema
O   Australia to Burma

Large horseshoe bat

Hipposideros dinops
O   Solomon Isl.

Leaf nosed bat

Hipposideros dyacorum
O   Borneo, (Sarawak)

Horseshoe bat

Hipposideros fuliginosus
O   Ghana to Cameroon

Sooty leaf nosed bat

Hipposideros fulvus
O   India, Afghanistan

Leaf nosed bat

Hipposideros galeritus
O   Australia, Sri Lanka

Fawn horseshoe bat

Hipposideros gentilis
O   Burma

Horseshoe bat

Hipposideros inexpectatus
O   Celebes

Leaf nosed bat

Hipposideros jonesi
O   W. Africa

Jones leaf nosed bat

Hipposideros lankadiva
O   Sri Lanka, India

Leaf nosed bat

Hipposideros larvatus
O   Burma to Viet Nam

Large roundleaf horseshoe bat

Hipposideros longicauda
O   Java

Horseshoe bat

Hipposideros lylei
O   Burma, Thailand, Malaysia

Shield faced bat

Hipposideros marisae
O   Guinea, Ivory Coast

Allens leaf nosed bat

Hipposideros muscinus
O   New Guinea

Leaf nosed bat

Hipposideros nequam
O   Selango

Malayan round leaf horseshoe bat

Hipposideros obscurus
O   Philippines

Leaf nosed bat

Hipposideros papua
O   New Guinea

Leaf nosed bat

Hipposideros pelingensis
O   Peleng Isl.

Leaf nosed bat

Hipposideros pomona
O   S.and S.E.Asia

Leaf nosed bat

Hipposideros pratti
O   S.China to Viet Nam

Pratt's leaf nosed bat

Hipposideros ridleyi
O   Singapore

Singapore roundleaf horseshoe bat (E)

Hipposideros ruber
O   W.Africa

Noacks African leaf nosed bat

Hipposideros sabanus
O   Sumatra, Borneo

Lawas roundleaf horseshoe bat

Hipposideros semoni
O   Australia

Warty nosed horseshoe bat

Hipposideros speoris
O   Sri Lanka, India

Schneider's leaf nosed bat

Hipposideros stenotis
O   Australia

Lesser warty nosed horseshoe bat

Hipposideros wollastoni
O   New Guinea

Leaf nosed bat

Paracoelops megalotis
O   Viet Nam ,Annam

Leaf nosed bat

Rhinonycteris arrantius
O   Australia

Orange leaf nosed bat

Triaenops furculus
O   Madagascar

Trouessarts trident bat

Triaenops humbloti
O   Madagascar

Humbolt's trident bat

Triaenops persicus
O   E. Africa, Iran

Persian leaf nosed bat

Triaenops rufus
O   Madagascar

Red Madagascan trident bat

FAMILY: PHYLLOSTOMATIDAE, New World Leaf nosed bats

Desmodus rotundus
O   Mexico to Chile, Argentina

Vampire bat

Diaemus youngi
O   Mexico to Brazil

White winged vampire bat

Diphylla ecaudata
O   Texas to Peru

Hairy legged vampire bat

Chrotopterus auritus
O   Mexico to Argentina

Peter's false vampire

Lonchorhina aurita
O   Chiapas to Peru, Bolivia

Tomes long eared bat

Macrophyllum macrophyllum
O   Guatemela to Peru

Large legged bat

Macrotus waterhousii
O   California to Guatemala

American leaf nosed bat

Micronycteris behni
O   Brazil, Peru, Bolivia

Small eared bat

Micronycteris brachyotis
O   French Guiana

Small eared bat

Micronycteris hirsuta
O   Costa Rica to Peru

Hairy small eared bat

Micronycteris megalotis
O   Mexico to Peru, Brazil

Small eared bat

Micronycteris minuta
O   Brazil, Colombia

Small eared bat

Micronycteris nicerfore
O   N.Colombia

Small eared bat

Micronycteris platyceps
O   Nicaragua to Venezuela

Small eared bat

Micronycteris pusilla
O   Brazil

Small eared bat

Micronycteris schimdtorum
O   Yucatan to W.Honduras

Small eared bat

Micronycteris sylvestris
O   Mexico to Peru,Brazil

Brown small eared bat

Mimon bennettii
O   Mexico to Brazil

Spear nosed bat

Mimon crenulatum
O   Guianas, Brazil, Peru

Spear nosed bat

Mimon koepckeae
O   Peru

Spear nosed bat

Phylloderma stenops
O   Honduras to Guianas

Spear nosed bat

Phyllostomus discolor
O   Chiapas to S.E.Brazil

Spear nosed bat

Phyllostomus elongatus
O   Ecuador, Guiana, Peru

Spear nosed bat

Phyllostomus hastatus   Spear nosed bat
O   Honduras to S.E.Brazil

Phyllostomus latifolius   Spear nosed bat
O   Guiana,(Kanaka Mts.)

Tonatia bidens   Spix round eared bat
O   Costa Rica to S. America

Tonatia brasiliensis   Round eared bat
O   Peru, Brazil

Tonatia carrikeri   Round eared bat
O   Venezuela,Bolivia,Peru

Tonatia minuta   Round eared bat
O   Venezuela, Trinidad

Tonatia sylvicola   Round eared bat
O   Belize to Bolivia, Brazil

Trachops cirrhosus   Fringe lipped bat
O   Guatemala to Brazil, Peru

Vampyrum spectrum   Linnaeus false vampire
O   Tabasco to Peru

Anoura brevirostrum   Tailless bat
O   Peru, Colombia

Anoura caudifera   Tailless bat
O   Colombia to Peru, Sao Paulo

Anoura cultrata   Long nosed bat
O   Costa Rica to Colombia

Anoura geoffroyi   Tailless bat
O   Mexico to Peru, Brazil

Glossophaga commissarisi   Long tongued bat
O   Mexico to Panama

Glossophaga elongata   Long tongued bat
O   Curacao Isl. Venezuela

Glossophaga longirostris   Long tongued bat
O   Lesser Antilles, Colombia

Glossophaga soricina   Long tongued bat
O   Mexico to Argentina

Lionycteris spurrelli   Long tongued bat
O   South America

Lonchophylla hesperia   Long tongued bat
O   Peru

Lonchophylla mordax   Long tongued bat
O   Panama to Brazil,Bolivia

Lonchophylla robusta   Long tongued bat
O   Panama to Peru

Lonchophylla thomasi   Long tongued bat
O   N.South America

Monophyllus plethodon   Long tongued bat
O   Lesser Antilles

Monophyllus rebmani   Long tongued bat
O   Greater Antilles

Platalina genovensium   Long nosed bat
O   Peru

Carollia brevicauda   Short tailed bat
O   Bolivia (Potosi)

Carollia castanea   Short tailed bat
O   Honduras to Bolivia

Carollia perspicillata (ssp)   Short tailed bat
O   S. America

Carollia subrufa   Short tailed bat
O   C.America to Peru

Rhinophylla alethina   Short tailed bat
O   Colombia

Rhinophylla fischerae   Short tailed bat
O   Peru,Brazil,Venezuela,Colombia

Rhinophylla pumilio   Short tailed bat
O   N.South America

Ametrida centurio   Tree bat
O   Amazon basin ,Brazil

Ametrida minor   Tree bat
O   Guianas

Ardops nichollsi   Antillean tree bat
O   Lesser Antilles

Ariteus flavescens   Jamaican fig eating bat
O   Jamaica

Artibeus aztecus   Fruit eating bat
O C.Mexico to Panama

Artibeus cinereus   Fruit eating bat
O Mexico to S.America

Artibeus glaucus   Fruit eating bat
O S.America

Artibeus hirsutus   Fruit eating bat
O Mexico

Artibeus inopinatus   Fruit eating bat
O Nicaragua, Honduras, El Salvador

Artibeus jamaicensis   Big fruit bat
O C. America to S. America

Artibeus lituratus   Fruit eating bat
O Mexico to S. America

Artibeus phaeotis   Fruit eating bat
O Yucatan to Peru

Artibeus quadrivittatus   Fruit eating bat
O S.America

Artibeus rosenbergi   Fruit eating bat
O S.America

Artibeus toltecus   Fruit eating bat
O Sinaloa to Panama

Artibeus watsoni   Fruit eating bat
O Yucatan to Panama,Colombia

Brachyphylla cavernarum West Indian fruit eating bat
O W. Indies

Brachyphylla nana   Fruit eating bat
O Cuba

Brachyphylla pumila   Fruit eating bat
O Hispanola

Centurio senex   Wrinkle faced bat
O Mexico to Venezuela

Chiroderma doriae   White lined bat
O E.Brazil

Chiroderma salvini   White lined bat
O Honduras to S.America

Chiroderma trinitatum   White lined bat
O   Panama, Trinidad, Peru

Chiroderma villosum   White lined bat
O   Mexico to Brazil

Choeroniscus godmani   Godmans bat
O   Chiapas, Mexico to Colombia

Choeroniscus inca   Long tongued bat
O   Guayana to Peru

Choeroniscus intermedius   Long tongued bat
O   Guayana , Venezuela, Peru

Choeronlscus minor   Long tongued bat
O   N.South America

Choeroniscus periosus   Long tongued bat
O   Colombia

Choeronycteris harrisoni   Banana bat
O   C.America

Choeronycteris mexicana   Mexican long tongued bat
O   S.W. United States to Guatemala

Ectophylla alba   Honduran white bat
O   Honduras, Nicaragua

Ectophylla macconnelli   White bat
O   Costa Rica to Bolivia

Enchisthenes harti   Little fruit eating bat
O   Mexico to N.E.South America

Hylonycteris underwoodi   Underwoods long tongued bat
O   Mexico to Panama

Leptonycteris curasoae   Long nosed bat
O   Antilles, Venezuela

Leptonycteris nivalis   Long nosed bat
O   S. Arizona, Texas to Nicaragua, Colombia

Leptonycteris sanborni   Long nosed bat
O   Texas to El Savador

Lichonycteris degener   Long nosed bat
O   Amazon basin, Brazil

Lichonycteris obscura
O   Guatemala to Peru

Brown long nosed bat

Phyllops falcatus
O   Cuba

Fig eating bat

Phyllops haitensis
O   Dominican Republic

Fig eating bat

Pygoderma bilabiatum
O   S.Mexico to S.E.Brazil

Ipanema bat

Scleronycteris ega
O   Amazonas, Brazil

Long nosed bat

Sphaeronycteris toxophyllum
O   N.South America

Fruit eating bat

Stenoderma rufum
O   Puerto Rico

Red fig eating bat

Sturnira aratathomasi
O   Colombia,Ecuador

Yellow shouldered bat

Sturnira bidens
O   Ecuador, Peru

Short tailed bat

Sturnira erythromos
O   Peru

Short tailed bat

Sturnira lilium
O   S.Mexico to Argentina

Yellow shouldered bat

Sturnira ludovici
O   S. Mexico to Peru

Anthony's bat

Sturnira magna
O   Amazonian region

Yellow shouldered bat

Sturnira mordax
O   Cartaga, Costa Rica

Hairy footed bat

Sturnira nana
O   Peru

Yellow shouldered bat

Sturnira thomasi
O   Guadeloupe

Yellow shouldered bat

Sturnira tildae
O   Trinidad to Peru

Yellow shouldered bat

Uroderma bilobatum
O   Yucatan to Bolivia

Tent making bat

Uroderma magnirostrum
O   Mexico to N.South America

Tent making bat

Vampyressa bidens
O   N.South America

Yellow eared bat

Vampyressa melissa
O   Peru

Yellow eared bat

Vampyressa nymphaea
O   Panama to S. America

Yellow eared bat

Vampyressa pusilla
O   Mexico to N.South America

Yellow eared bat

Vampyressa thyone
O   Costa Rica to Ecuador

Yellow eared bat

Vampyrodes caraccioloi
O   Honduras to N.Brazil

San Pablo bat

Vampyrops brachycephalus
O   Colombia, Venezuela,Peru

Broad nosed bat

Vampyrops dorsalis
O   Venezuela to Peru

Broad nosed bat

Vampyrops helleri
O   Oaxaca to S.America

Broad nosed bat

Vampyrops intermedius
O   Colombia

Broad nosed bat

Vampyrops lineatus
O   Paraguay, Brazil.

Broad nosed bat

Vampyrops nigellus
O   Colombia to Peru

Broad nosed bat

Vampyrops vittatus
O   Costa Rica to Peru

White lined bat

Erophylla bombifrons
O   Puerto Rico, Dominican Republic

Flower bat

Erophylla sezekorni
O   Cuba, Jamaica, Bahamas

Buffy flower bat

Phyllonycteris aphylla
O   Jamaica

Flower bat

Phyllonycteris major
O   Puerto Rico

Flower bat

Phyllonycteris obtusa   Flower bat
O   Haiti

Phyllonycteris poeyi   Flower bat
O   Cuba

FAMILY: MORMOOPIDAE,Moustache bats

Mormoops blainvillii   Antillean ghost faced bat
O   Greater Antilles

Mormoops megalophylla   Ghost faced bat
O   Texas to C. America, Colombia

Pteronotus davyi   Davy's naked backed bat
O   Mexico to N.South America

Pteronotus fuliginosus   Sooty moustached bat
O   Greater Antilles

Pteronotus macleayii   Mac Leay's moustached bat
O   Cuba, Jamaica

Pteronotus parnelli   Moustached bat
O   Greater Antilles, Mexico to S. America

Pteronotus personatus   Moustached bat
O   Mexico to Colombia

Pteronotus suapurensis   Big naked backed bat
O   Guatemala to Venezuela

FAMILY: NATALIDAE, Funnel eared bats

Natalus lepidus   Long legged bat
O   Cuba

Natalus major   Funnel eared bat
O   Dominican Republic

Natalus micropus   Long legged bat
O   Jamaica, Colombia

Natalus stramineus   Funnel eared bat
O   E.Brazil to Mexico

Natalus tumidirostris   Long legged bat
O   Bahamas, Venezuela, Colombia

FAMILY: FURIPTEDIDAE, Thumbless bats

Amorphochilus schnablii   Smoky bat
O   Peru, Ecuador

THUMBLESS BAT, NEW WORLD AND OLD WORLD DISC WINGED BATS

Furipterus horrens   Smoky bat
O   Costa Rica to N.South America

FAMILY: THYROPTERIDAE, New World disc winged bats

Thyroptera discifera   Disc winged bat
O   Honduras to Peru

Thyroptera tricolor   Disc winged bat
O   Belize to Peru

FAMILY : MYZOPODIDAE, Old World disc winged bats

Myzopoda aurita   Sucker footed bat
O   Madagascar

FAMILY: VESPERTILIONIDAE, Plain nosed bats

Baeodon alleni   Allen's baeodon
O   Mexico,(Jalisco to Oaxaxa)

Barbastella barbastellus   Common barbastrelle
O   Morocco, Europe, W.Asia

Barbastella leucomelas   Eastern barbastrelle
O   Sinai, S.E.Asia

Chalinolobus alboguttatus   Allen's striped bat
O   Zaire

Chalinolobus argentata   Silvered bat
O   C.Africa

Chalinolobus beatrix   Beatrix bat
O   Gabon to Uganda

Chalinolobus dwyeri   Large eared pied bat
O   Australia

Chalinolobus egeria   Bibundi bat
O   Cameroon

Chalinolobus gouldii   Lobe lipped bat
O   Australia

Chalinolobus morio   Chocolate bat
O   Australia

Chalinolobus nigrogriseus   Pied bat
O   Australia, New Guinea

Chalinolobus picatus   Little pied bat
O   Australia, New Guinea

Chalinolobus poensis   Abo bat
O   W.Africa to Congo

Chalinolobus superba   Pied bat
O   Ghana

Chalinolobus tuberculatus   Pied bat
O   New Zealand

Chalinolobus variegatus   Butterfly bat
O   Africa

Eptesicus andinus   Big brown bat
O   Mexico to Peru

Eptesicus bobrinskoi   Borrinsky's bat
O   Turkmen, Iran

Eptesicus bottae   Bottas serotine
O   Egypt, S.E.Europe, S.W.Asia

Eptesicus brasiliensis   Brown bat
O   Mexico to Argentina

Eptesicus brunneus   Dark brown bat
O   Nigeria to Zaire

Eptesicus capensis   Cape serotine
O   Africa

Eptesicus chiriquinus   Brown serotine
O   Panama

Eptesicus diminutus   Brown serotine
O   Brazil,(Bahia)

Eptesicus fidelis   Brown serotine
O   N.W.Argentina to E.Brazil

Eptesicus floweri   Lesser Sudan horn skinned bat
O   Sudan

Eptesicus furinalis   Brown bat
O   Nicaragua to Argentina, Paraguay

Eptesicus fuscus   Big brown bat
O   S.Canada to N.South America

Eptesicus guineensis   Tiny serotine
O   W.Africa to Zaire

Eptesicus hottentotus   Long tailed house bat
O   Namibia, Zambia,South Africa

Eptesicus innoxius   Brown bat
O   E.Ecuador, Peru

Eptesicus loveni   Loven's serotine
O   W.Kenya

Eptesicus melanopterus   Brown bat
O   N.E.Brazil, Guayanas

Eptesicus melckorum   Melck's house bat
O   Zambia,South Africa

Eptesicus montosus   Brown bat
O   N.South America

Eptesicus nasutus   Sind bat
O   Arabia to India

Eptesicus nilssoni   Bat
O   Europe, Russia, Kashmir

Eptesicus notius   Cape horn skinned bat
O   Cape Town

Eptesicus pachyotis   Thick eared bat
O   Assam, India

Eptesicus platyops   Lagos serotine
O   Nigeria, Senegal

Eptesicus pumilus   Little bat
O   Australia

Eptesicus rendalli   Rendall's serotine
O   Gambia to Mozambique

Eptesicus serotinus   Serotine
O   N.Africa, Europe, Asia

Eptesicus sodalis   Serotine
O   Rumania to Mongolia

Eptesicus somalicus   Somali serotine
O   Africa

Eptesicus tenuipinnis   White winged serotine
O   Guinea to Tanzania

Eptesicus walli   Wall's serotine
O   Iraq,Iran

Eptesicus zuluensis   Alde bat
O   Africa

Euderma maculatum   Spotted bat
O   W.United States to Mexico

Eudiscopus denticulus   Disc footed bat
O   S.Asia, Philippines

Glischropus tylopus   Thick thumbed pipistrelle
O   Burma to Borneo

Hesperoptenus blanfordi   Lesser false serotine
O   S.Burma to Malaysia

Hesperoptenus doriae   False serotine
O   Borneo

Hesperoptenus tickelli   Tickell's bat
O   India, Sri Lanka, Thailand

Hesperoptenus tomesi   False serotine bat
O   Malaysia

Histiotus alienus   Big eared brown bat
O   E.Brazil

Histiotus macrotus   Big eared brown bat
O   Bolivia, Argentina, Peru

Histiotus montanus   Big eared brown bat
O   Peru to Tierra Del Fuego

Histiotus velatus   Big eared brown bat
O   Brazil

Laephotis wintoni   De Winton's long eared bat
O   Angola to Kenya

Lasionycteris noctivagans   Silver haired bat
O   Alaska to Mexico

Lasiurus boeralis   Red bat
O   S.Canada to Argentina

Lasiurus brachyotis   Hairy tailed bat
O   Galapagos Isl.

Lasiurus castaneus   Hairy tailed bat
O   Panama

Lasiurus cinereus   Hoary bat
O   N.America to S. America, Hawaii, Iceland

Lasiurus degelidus   Hairy tailed bat
O   Jamaica

Lasiurus ega   Western yellow bat
O   S.W United States to Argentina

Lasiurus intermedius   Eastern yellow bat
O   E.United States, Mexico, Cuba

Lasiurus seminolus   Seminole bat
O   S.W. United States to Vera Cruz, Mexico

Mimetillus moloneyi   Moloney's flat headed bat
O   W.to E. Africa

Myotis adversus   Large footed myotis
O   Australia to Thailand

Myotis albescens   Paraguay myotis
O   Costa Rica to Argentina, Chile

Myotis altarium   Bat
O   China,(Szechuan)

Myotis annectans   Bat
O   India to Thailand

Myotis argentatus   Bat
O   Mexico.(Vera Cruz)

Myotis auriculus   Bat
O   Mexico

Myotis australis   Small footed myotis
O   Nsw.

Myotis austroriparius   Mississippi myotis
O   S.E. United States

Myotis bartelsi   Mouse eared bat
O   Java

Myotis bechsteinii   Bechstein's bat
O   Europe, Russia

Myotis blythii   Lesser mouse eared bat
O   N.Africa, Europe,Asia

Myotis bocagei   Rufous mouse eared bat
O   Liberia to Tanzania

Myotis browni   Bat
O   Philippines,(Mindanao)

Myotis californicus   California myotis
O   W.N.America to Mexico

Myotis capaccinii   Long fingered bat
O   N.Africa, S.Europe, S.W.Asia

Myotis chiloensis   Mouse eared bat
O   Chile, N.W.Argentina to Ecuador

Myotis dasycneme   Pond bat
O   N.Europe to W.Siberia

Myotis daubentoni   Daubenton's bat
O   Europe to Japan

Myotis davidii   Bat
O   China

Myotis elegans   Bat
O   Mexico to Costa Rica

Myotis emarginatus   Geoffroy's bat
O   Algeria, Europe, S.W.Asia

Myotis evotis   Long eared myotis
O   S.W.Canada to C. Mexico

Myotis formosus   Hodgsons bat
O   China, Korea, Nepal,Japan

Myotis fortidens   Bat
O   Mexico

Myotis frater   Bat
O   China, Korea, E.Siberia

Myotis goudoti   Malagasy mouse eared bat
O   Madagascar

Myotis grisescens   Gray myotis (E)
O   Kansas to Virginia and Florida

Myotis hasseltii   Lesser large footed bat
O   Malaysia, Java

Myotis hermani   Mouse eared bat
O   Sumatra

Myotis herrei   Bat
O   Philippines,(Luzon)

Myotis horsfieldii   Horsfields bat
O   Malaysia, Java

Myotis jeannei   Bat
O   Philippines,(Mindanao)

Myotis keaysi   Bat
O   Mexico, (Yucatan) to Peru

Myotis keenii   Keens myotis
O   United States ,Canada

Myotis laniger   Mouse eared bat
O   China

Myotis leibii   Small footed myotis
O   United States

Myotis lesueuri   Wing gland bat
O   South Africa

Myotis longipes   Mouse eared bat
O   Kasmir, Afghanistan

Myotis lucifugus   Little brown myotis
O   Alaska to N. Mexico

Myotis macrodactylus   Bat
O   Japan

Myotis montivagus   Burmese whiskered bat
O   India, S.China

Myotis morrisi   Bat
O   Ethiopia

Myotis myotis   Large mouse eared bat
O   Europe, China

Myotis mystacinus   Whiskered bat
O   Palaearctic

Myotis nattereri   Natterer's bat
O   Morocco, Europe, Asia

Myotis nigricans   Black myotis
O   Mexico to Argentina

Myotis oreias   Singapore whiskered bat
O   Singapore

Myotis patriclae   Bat
O   Philippines

Myotis pequinius   Bat
O   China

Myotis peytoni   Mouse eared bat
O   Malaysia

Myotis ricketti   Big footed bat
O   China

Myotis ruber   Mouse eared bat
O   Paraguay,Brazil to Uruguay

Myotis rufopictus   Bat
O   Philippines

Myotis scotti   Scott's mouse eared bat
O   Ethiopia

Myotis seabrai   Lesuer's wing gland bat
O   South Africa

Myotis sicarius   Bat
O   India

Myotis siligorensis   Whiskered bat
O   Burma, Thailand

Myotis simus   Mouse eared bat
O   Nicaragua to Ecuador, Brazil

Myotis sodalis   Indiana myotis (V)
O   E. United States

Myotis stalkeri   Bat
O   Kei Isl.

Myotis thysanodes   Fringed myotis
O   W.North America to Mexico

Myotis tricolor   Cape hairy bat
O   S. Africa to Ethiopia

Myotis velifer   Cave myotis
O   S.W. United States to Honduras

Myotis volans   Long legged myotis
O   Alaska to Mexico

Myotis weberi   Bat
O   S.Celebes

Myotis welwitschii   Welwitsch's bat
O   S.Africa to Ethiopia

Myotis yumanensis   Yuma myotis
O   W.North America to Mexico

Nyctalus azoreum   Noctule
O   Azores Isl.

Nyctalus joffrei
O   Burma

Noctule

Nyctalus lasiopterus
O   W.Europe

Giant noctule

Nyctalus leisleri
O   Europe to W.Himalayas

Lesser noctule

Nyctalus montanus
O   N.India

Noctule

Nyctalus noctula
O   Morocco, Europe,Asia

Common noctule

Nyctalus stenopterus
O   Malaysia, Borneo

Malaysian noctule

Nyctalus verrucosus
O   Madeira

Madeiran noctule

Nycticeius albofuscus
O   Senegal to Tanzania

Light winged lesser house bat

Nycticeius cubanus
O   Cuba

Evening bat

Nycticeius emarginatus
O   India

Large eared yellow bat

Nycticeius greyi
O   Australia, Papua

Little broad nosed bat

Nycticeius hirundo
O   Ghana to Tanzania

Dark winged lesser house bat

Nycticeius humeralis
O   E.United States to E.Mexico

Evening bat

Nycticeius inflatus
O   Qld.

Hughenden broad nosed bat

Nycticeius pallidus
O   India

Yellow desert bat

Nycticeius rueppellii
O   Australia

Greater broad nosed bat

Nycticeius sanborni
O   Papua

Evening bat

Nycticeius schlieffeni
O   Africa to S.W.Arabia

Schlieffen's bat

Otonycteris hemprichi
O   N.Africa to Iraq

Hemprich's big eared bat

Philetor rohui
O   New Guinea, Malaysia

New Guinea brown bat

Pipistrellus aero
O   N.W.Kenya

Mt Gargues pipistrelle

Pipistrellus abramus
O   Siberia, Japan, China, S.E.Asia

Japanese pipistrelle

Pipistrellus affinis
O   China

Chocolate bat

Pipistrellus anchietai
O   Angola,Zaire, Zambia

Anchietas pipistrelle

Pipistrellus angulatus
O   New Guinea, Solomons

Pipistrelle

Pipistrellus ariel
O   Egypt, Sudan

Desert pipistrelle

Pipistrellus babu
O   India

Pipistrelle

Pipistrellus brachypterus
O   Sumatra

Pipistrelle

Pipistrellus cadornae
O   Thailand, Burma

Pipistrelle

Pipistrellus ceylonicus
O   Sri Lanka to Borneo

Kelaart's pipistrelle

Pipistrellus circumdatus
O   Burma to Java

Large black pipistrelle

Pipistrellus coromandra
O   India to S.China

Indian pipistrelle

Pipistrellus crassulus
O   Cameroon, Congo,Zaire

Broad headed pipistrelle

Pipistrellus deserti
O   Libya, Algeria

Desert pipistrelle

Pipistrellus dormeri
O   India, Taiwan

Dormer's bat

Pipistrellus eisentrauti
O   Cameroon,Ivory Coast

Eisentraut's pipistrelle

Pipistrellus endoi   Bat
O   Japan

Pipistrellus hesperus   Western pipistrelle
O   W. United States to Mexico

Pipistrellus imbricatus   Brown pipistrelle
O   Malaysia Peninsula ,Java

Pipistrellus inexspectatus   Allen's pipistrelle
O   Cameroon,Zaire

Pipistrellus io   Great pipistrelle
O   Philippines, (Luzon)

Pipistrellus javanicus   Yellow headed pipistrelle
O   Australia to Japan

Pipistrellus kuhli   Kuhl's pipistrelle
O   Africa, Europe, Asia

Pipistrellus lophurus   Pipistrelle
O   S.Burma

Pipistrellus maderensis   Madeira pipistrelle
O   Madeira

Pipistrellus mimus   India pygmy pipistrelle
O   Sri Lanka, India, Burma

Pipistrellus minahassae   Pipistrelle
O   Celebes

Pipistrellus mordax   Pipistrelle
O   Sri Lanka, India, Java

Pipistrellus murrayi   Pipistrelle
O   Christmas Isl.

Pipistrellus musciculus   Least pipistrelle
O   Cameroon to Zaire

Pipistrellus nanulus   Tiny pipistrelle
O   Nigeria to Zaire

Pipistrellus nanus   Banana bat
O   Africa, Madagascar

Pipistrellus nathussi   Pipistrelle
O   Europe, W.Asia Minor

Pipistrellus papuanus   Papuan pipistrelle
O   Australia, New Guinea

Pipistrellus peguensis
O   Burma
Pipistrelle

Pipistrellus permixtus
O   Tanzania
Dar Es Salaam pipistrelle

Pipistrellus petersi
O   N.Celebes
Pipistrelle

Pipistrellus pipistrellus
O   Morocco, Europe,Asia
Common pipistrelle

Pipistrellus pulveratus
O   S. China
Pipistrelle

Pipistrellus ridleyi
O   Malaysia
Pipistrelle

Pipistrellus rosseti
O   Cambodia
Thick thumb bat

Pipistrellus rueppelli
O   Africa to Iraq
Ruppell's bat

Pipistrellus rusticus
O   Africa
Rusty bat

Pipistrellus savii
O   Africa, Europe, Asia
Savis pipistrelle

Pipistrellus societatis
O   Malaysia
Pipistrelle

Pipistrellus subflavus
O   E, United States to Honduras
Eastern pipistrelle

Pipistrellus tasmaniensis
O   Australia, Tasmania
Tasmanian pipistrelle

Pipistrellus tenuis
O   Malaysia, Borneo,Australia
Least pipistrelle

Pipistrellus tralatitius
O   Malaysia, Sumatra, Borneo
Pipistrelle

Plecotus auritus
O   Europe, Russia, China
Long eared bat

Plecotus austriacus
O   Africa, Europe,China
Gray long eared bat

Plecotus phyllotis
O   E.United States
Allen's big eared bat

Plecotus rafinesquii   Big eared bat
O   S.E. United States

Plecotus townsendii   Big eared bat
O   S.W.Canada to Mexico

Rhogeessa gracilis   Slender yellow bat
O   Mexico (Pueblo)

Rhogeessa minutilla   Little yellow bat
O   Venezuela,Colombia

Rhogeessa mira   Little yellow bat
O   Mexico, (Michoacan)

Rhogeessa parvula   Little yellow bat
O   Mexico to S. America

Rhogeessa tumida   Little yellow bat
O   Mexico to Venezuela, Colombia

Scotomanes ornatus   Harlequin bat
O   S.Asia

Scotophilus celebensis   Greater yellow bat
O   Celebes

Scotophilus gigas   Great Brown bat
O   W.and C.Africa

Scotophilus heathi   Greater yellow bat
O   India to Malaysia

Scotophilus kuhlii   Brown bat
O   S.Asia

Scotophilus leucogaster   Lesser yellow house bat
O   Africa

Scotophilus nigrita   Yellow house bat
O   Africa

Scotophilus temmincki   House bat
O   Sri Lanka to Philippines,Java

Tylonycteris pachypus   Club footed bat
O   India to Philippines, Java

Tylonycteris robustula   Greater flat headed bat
O   S.E.Asia

Vespertilio murinus   Particolored bat
O   Europe, Russia to Afghanistan

9 1

Vespertilio superans   Bat
O   Siberia, Japan, Korea, S.China

Miniopterus australis   Little bent wing bat
O   Australia to Borneo

Miniopterus fraterculus   Lesser long fingered bat
O   S. Africa

Miniopterus inflatus   Greater long fingered bat
O   Cameroon to E.Africa

Miniopterus medius   S.E.Asian bent winged bat
O   S.E.Asia,Indonesia

Miniopterus minor   Least long fingered bat
O   Zaire, Tanzania, Madagascar

Miniopterus schriebersi   Long winged bat
O   Africa, Europe, Asia, Australia

Miniopterus tristis   Long fingered bat
O   New Guinea, Philippines

Harpiocephalus harpia   Hairy winged bat
O   S.Asia, Java, Sumatra

Murina aenea   Bronzed tube nosed bat
O   Malaysia

Murina aurata   Little tube nosed bat
O   S.E.Siberia, Japan

Murina balstoni   Tube nosed bat
O   Java

Murina cyclotis   Round eared tube nosed bat
O   Sri Lanka to S. China

Murina florium   Tube nosed bat
O   Ceram, Buru

Murina grisea   Peters tube nosed bat
O   India

Murina huttoni   Hutton tube nosed bat
O   India to Malaysia

Murina leucogaster   Great tube nosed bat
O   Siberia, China, Japan

Murina suilla   Tube nosed bat
O   Malaysia, Borneo

Murina tenebrosa
O   Japan

Tube nosed bat

Murina tubinaris
O   Viet Nam, Laos,Burma

Tube nosed bat

Kerivoula africana
O   Tanzania

Tanzanian woolly bat

Kerivoula agnella
O   New Guinea

Painted bat

Kerivoula argentata
O   Africa

Damara woolly bat

Kerivoula bombifrons
O   Borneo, Java

Woolly bat

Kerivoula cuprosa
O   Cameroon,Kenya

Copper woolly bat

Kerivoula hardwickii
O   Sri Lanka to Philippines

Painted bat

Kerivoula harrisoni
O   Africa

Harrison's woolly bat

Kerivoula javana
O   Borneo

Groove toothed bat

Kerivoula lanosa
O   Africa

Lesser woolly bat

Kerivoula minuta
O   Thailand, Malaysia

Least forest bat

Kerivoula muscilla
O   W.Africa

Bates woolly bat

Kerivoula muscina
O   Papua

Painted bat

Kerivoula myrella
O   Admiralty Isl.

Painted bat

Kerivoula papillosa
O   India to Borneo

Papillose bat

Kerivoula papuensis
O   Qld. Papua

Dome headed bat

Kerivoula pellucida
O   Sumatra to Philippines

Clear winged bat

Kerivoula phalaena
O   Liberia to Congo

Spurrell's woolly bat

Kerivoula picta
O   Sri Lanka to Borneo

Painted bat

Kerivoula pusilla
O   Borneo,(Sarawak)

Woolly bat

Kerivoula rapax
O   Celebes

Painted bat

Kerivoula smithi
O   Nigeria to Kenya

Smith's woolly bat

Kerivoula whiteheadi
O   Thailand, Borneo, Philippines

Little forest bat

Antrozous dubiaquercus
O   Mexico, (Tres Maras Isl)

Pallid bat

Antrozous koopmani
O   Cuba

Pallid bat

Antrozous pallidus
O   W. of North America to Mexico

Pallid bat

Laningtona lophorhina
O   Papua

Bat

Nyctophilus arnhemensis
O   Nt.

Arnheim land long eared bat

Nyctophilus bifax
O   Nt.,Qld.

N.Queensland long eared bat

Nyctophilus geoffroyi
O   Australia,Tasmania

Lesser long eared bat

Nyctophilus microdon
O   New Guinea

Big eared bat

Nyctophilus microtis
O   Papua

Big eared bat

Nyctophilus timoriensis
O   Australia, Timor

Greater long eared bat

Nyctophilus walkeri
O   Nt.

Big eared bat

Pharotis imogene
O   New Guinea

Papuan big eared bat

Tomopeas ravus   Bat
O   Peru

FAMILY: MYSTACINIDAE,New Zealand Short tailed bats

Mystacina tuberculata   New Zealand short tailed bat
O   New Zealand

FAMILY: MOLOSSIDAE, Free tailed bats

Cheiromeles parvidens   Hairless bat
O   Celebes,Philippines

Cheiromeles torquatus   Hairless bat
O   Thailand to Java ,Borneo

Eumops abrasus   Temminck's mastiff bat
O   S.Mexico to Bolivia

Eumops amazonicus   Mastiff bat
O   Brazil

Eumops auripendulus   Mastiff bat
O   Mexico to Peru

Eumops bonariensis   Peter's mastiff bat
O   Mexico to Argentina

Eumops glaucinus   Mastiff bat
O   Florida,Mexico to S. America, Cuba

Eumops hansae   Mastiff bat
O   Costa Rica to N.South America

Eumops maurus   Guianan mastiff bat
O   Guianas, Surinam

Eumops perotis   Greater mastiff bat
O   S.California to Mexico, Peru

Eumops trumbulli   Mastiff bat
O   Colombia to N.Peru

Eumops underwoodi   Mastiff bat
O   Arizona to Honduras

Molossops aequatorianus   Dog faced bat
O   W. Ecuador

Molossops brachymeles   Dog faced bat
O   South America

Molossops greenhalli   Dog faced bat
O Trinidad,Mexico to Surinam

Molossops milleri   Dog faced bat
O E.Peru,(Amazon)

Molossops planirostris   Dog faced bat
O Panama to Paraguay, Brazil

Molossops temminckii   Dog faced bat
O Colombia to Argentina

Molossus ater   Mastiff bat
O Mexico to Argentina

Molossus barnesi   Mastiff bat
O Brazil , French Guiana

Molossus bondae   Free tailed bat
O Nicaragua to Colombia

Molossus molossus   Free tailed bat
O Mexico to Argentina

Molossus sinaloae   Free tailed bat
O Mexico to Venezuela

Molossus tropidorhynchus Mastiff bat
O Cuba

Myopterus albatus   Mastiff bat
O Ivory Coast, Zaire

Myopterus daubentonii   Mastiff bat
O Senegal

Myopterus whitleyi   Mastiff bat
O W.Central Africa

Neoplatymops mattogrossensis Flat headed bat
O Venezuela, Brazil

Otomops formosus   Free tailed bat
O Java

Otomops martiensseni   Martienssens free tailed bat
O E.and C. Africa

Otomops papuensis   Free tailed bat
O New Guinea

Otomops secundus   Free tailed bat
O New Guinea

Otomops wroughtoni   Free tailed bat
O   S.India

Platymops setiger   Flat headed free tailed bat
O   Kenya,Ethiopia

Promops centralis   Mastiff bat
O   Mexico to Paraguay

Promops nasutus   Dome Palate mastiff bat
O   Brazil, Argentina,Surinam

Promops pamana   Mastiff bat
O   Brazil

Sauromys petropohilus   Flat headed bat
O   S.Africa

Tadarida acetabulosus   Natal wrinkle lipped bat
O   South Africa,(Natal), Ethiopia

Tadarida aegyptiaca   Egyptian free tailed bat
O   Africa, India

Tadarida africana   Giant African free tailed bat
O   Ethiopia

Tadarida aloysiisabaudiae   Duke of Abruzzis free tailed bat
O   Uganda,Zaire,Gabon

Tadarida ansorgei   Ansorge's free tailed bat
O   Cameroon to Tanzania

Tadarida aurispinosa   Free tailed bat
O   Mexico to N.South America

Tadarida australis   White striped bat
O   Australia, New Guinea

Tadarida beccarii   Free tailed bat
O   Amboina, New Guinea

Tadarida bemmeleni   Gland tailed bat
O   E.and W.Africa

Tadarida bivittata   Spotted free tailed bat
O   Ethiopia to Zambia

Tadarida brasillensis   Mexico free tailed bat
O   S. United States to Argentina

Tadarida chapini   Long crested free tailed bat
O   Senegal to Congo

Tadarida condylura
O   Africa

Angola free tailed bat

Tadarida congica
O   Uganda, Zaire

Medje greater free tailed bat

Tadarida demonstrator
O   Sudan, Uganda,Zaire

Mongalla free tailed bat

Tadarida femorosacca
O   S. United States to Mexico

Pocketed free tailed bat

Tadarida fulminans
O   E.Africa

Madagascar free tailed bat

Tadarida gracilis
O   Brazil (Mato Grosso)

Free tailed bat

Tadarida jobensis
O   Australia, New Guinea

Northern masstiff bat

Tadarida johorensis
O   Malaysia, Sumatra

Dato Meldrums bat

Tadarida jugularis
O   Madagascar

Wrinkle lipped bat

Tadarida laticaudata
O   Mexico to S.Brazil

Broad tailed bat

Tadarida leonis
O   Sierre Leone to Zaire

Sierra Leone free tailed bat

Tadarida lobata
O   N. Kenya

Big eared free tailed bat

Tadarida loriae
O   Australia,New Guinea

Little scurrying bat

Tadarida macrotis
O   Nevada to S. America

Big free tailed bat

Tadarida major
O   Mali to Tanzania

Lappet eared free tailed bat

Tadarida midas
O   Africa, Madagascar

Midas bat

Tadarida molossus
O   Brazil to Surinam, Peru

Free tailed bat

Tadarida mops
O   Malaysia, Sumatra, Borneo

Free tailed bat

Tadarida nanula
O   W. Africa to Kenya

Dearf free tailed bat

Tadarida nigeriae
O   Nigeria to Zimbabwe

Nigerian free tailed bat

Tadarida niveiventer
O   Zaire, Tanzania

White bellied free tailed bat

Tadarida norfolkensis
O   Norfolk Isl, (S.W.Pacific)

Norfolk Isl. scurrying bat

Tadarida planiceps
O   Australia

Little flat bat

Tadarida plicata
O   Sri Lanka to Borneo

Wrinkled lipped bat

Tadarida pumila
O   Africa,Madagascar

Little free tailed bat

Tadarida russata
O   Zaire,Cameroon

Russer free tailed bat

Tadarida sarasinorum
O   Celebes

Free tailed bat

Tadarida similis
O   Colombia, Peru

Free tailed bat

Tadarida teniotis
O   N. Africa, Europe

European free tailed bat

Tadarida thersites
O   W.Africa to Zaire

Railed bat

Tadarida tragata
O   Sri Lanka, India

Free tailed bat

Tadarida trevori
O   Zaire,Uganda

Niangara free tailed bat

Tadarida yucatanica
O   Yucatan Peninsula, Guatemala

Free tailed bat

Xiphonycteris spurrelli
O   Ghana, Togo,Zaire

Spurrell's free tailed bat

TOTAL CHIROPTERA : O	Z	W	P.W	P.Z

_____

_____

_____

_____

_____

_____

_____

_____

_____

_____

_____

_____

_____

_____

_____

_____

_____

_____

_____

_____

_____

_____

ORDER: CHIROPTERA

ORDER: CHIROPTERA

ORDER:PRIMATES

This very diverse order is divided into three suborders:Prosimii, the more primitive, includes Lemurs, Indris, Lorises, and Ayes-ayes. Tarsioidea, possibly an intermediary,contains only one animal, the Tarsier. Anthropoidea, the more advanced, contains Marmosets, Monkeys, Apes , and Man. In the living families, we see a trend of characteristics that may have led to Man, however, there is no direct descent from one family to another.

The earliest primates were tree dwelling animals, and most are still arboreal today. Primates exhibit many adaptations for arboreal life. The skeleton is very generalized with much flexibility. Locomotion is usually quadrupedal, but there is a trend toward an upright stance which frees the hands for feeding.

Most primates climb by grasping, which reduces the risk of falling. Some primates have become quite large for arboreal animals, since grasping allows them to spread their weight among several branches. Grasping began with the development of long, spreadable fingers and toes. Most primates retain five fingers and toes, usually with nails instead of claws . Most have a functionally opposable thumb, which allows them to use the hand not only for grasping, but for procuring food and grooming, Most have an opposable great toe.

Good vision is important for efficient movement in the trees. In some of the primitive primates, the eyes are not completely in the front of the head, and the muzzle is elongated but as we move up the "Evolutionary ladder", we find that in higher primates the eyes are in the front. The area seen by one eye overlaps part of that seen by the other, giving the animal three dimensional (Stereoscopic) vision. The face is flattened,the tooth row shortened. The teeth are heterodont (of varied types) for a primarily omnivorous diet. The muzzle is reduced and therefore the sense of smell is not as acute.

A very important trend in primates, is that of the greater development of the brain. As we go from primitive to more advanced primates, we find the brain becoming more globular in shape, the cerebral hemispheres expanded, culminating with the very large brain of Man.

Primates are found worldwide, but are most successful in tropical and subtropical areas. Although mostly arboreal, some have become partly or fully terrestrial (Baboons, Chimpanzee, Man) .Most live in social groups which are essential to their survival. Group life allows the young ample time for maturation and also gives greater protection from predators.

SUBORDER: PROSIMII

FAMILY: LEMURIDAE, Lemurs

Lemurs are heavily furred, long-tailed primates, which range in size from a mouse to a cat and inhabit only the island of Madagascar and the Comoro Islands. They are endangered mainly due to the destruction of their habitat. Their survival is probably due to their isolation as they do not live in association with more progressive primates.

They exhibit many primitive (for primates) characteristics, including a long narrow skull, long moist muzzle, and facial vibrissae. Vision is not as well developed as in more advanced primates.

Many lemurs are arboreal, climbing by grasping. The thumb and big toe are somewhat opposable.The whole hand is used in grasping and food handling. Some Lemurs live in social groups with well defined dominance hierarchies.

Hapalemur griseus		Gray gentle lemur (V)	
O Madagascar			
Z	W		P.W
			P.Z
Hapalemur simus		Broad nosed gentle lemur (R)	
O S.E. Madagascar			
Z	W		P.W
			P.Z
Lemur catta		Ring tailed lemur (E*)	
O S.E. to S.W. Madagascar			
Z	W		P.W
			P.Z
Lemur coronatus		Crowned lemur	
O N.Madagascar			
Z	W		P.W
			P.Z
Lemur fulvus		Brown lemur	
O Madagascar			
Z	W		P.W
			P.Z
Lemur macaco		Black lemur (E)	
O Madagascar			
Z	W		P.W
			P.Z
Lemur mongoz		Mongoose lemur (V)	
O N.W.Madagascar			
Z	W		P.W
			P.Z
Lemur rubriventer		Red bellied lemur (E*)	
O Madagascar			
Z	W		P.W
			P.Z
Lemur variegatus		Ruffed lemur (E*)	
O Madagascar			
Z	W		P.W
			P.Z

Lepilemur mustelinus		Weasel lemur (V)	
O Madagascar			
Z	W		P.W
			P.Z

Cheirogaleus furcifer		Fork marked mouse lemur (I)	
O N.and N.W. Madagascar			
Z	W		P.W
			P.Z

Cheirogaleus major		Greater dwarf lemur (E*)	
O Madagascar			
Z	W		P.W
			P.Z

Cheirogaleus medius		Fat tailed lemur (V)	
O S.E. Madagascar			
Z	W		P.W
			P.Z

Cheirogaleus trichotis		Dwarf lemur (R)	
O E.Madagascar			
Z	W		P.W
			P.Z

Microcebus coquereli		Coquerels mouse lemur (V)	
O Madagascar			
Z	W		P.W
			P.Z

Microcebus murinus		Lesser mouse lemur (E*)	
O Madagascar			
Z	W		P.W
			P.Z

Microcebus rufus		Brown lesser mouse lemur	
O E.Madagascar			
Z	W		P.W
			P.Z

FAMILY: INDRIDAE, Woolly Indris,   Sifaka,   Indris

These rather large primates are restricted to Madagascar. They have long hind limbs. The hands and feet are well adapted for grasping. They climb rather slowly and cling to tree trunks in a vertical position. On the ground they stand upright and move by a series of leaps or hops. Primarily herbivorous, the Indris occupy the same niche as do Langurs and Colobus of Asia and Africa and Howlers of South America. Members of the genus Indri, like the howlers give loud resonant calls,especially in the morning and evening, perhaps to maintain territorial boundaries.

Avahi laniger		Woolly indri (V)	
O Madagascar			
Z	W		P.W
			P.Z

Indri indri		Indris (E)	
O N.E.Madagascar			
Z	W		P.W
			P.Z

Propithecus diadema   Diademed sifaka (E*)
O   Madagascar
          Z   W   P.W
                                                                P.Z
Propithecus verreauxi   Verreaus's sifaka (E*)
O   Madagascar
          Z   W   P.W
                                                                P.Z

FAMILY: DAUBENTONIIDAE, Aye-Aye

The aberrant aye-aye is a small,nocturnal primate which lives in dense forests and stands of bamboo in northern Madagascar.The skull is rather short; the dentition is quite different from that of most primates. The incisors are large, curved, and ever growing- - similar to those of rodents- - with enamel on the front surface only. There is a large diastema (gap) between the incisors and the premolars. An upper canine is sometimes present.The hand is also unique with long, slender, clawed digits. The third digit is remarkably slender and almost "wirelike".The thumb is not opposable, but the great toe is and is the only digit to bear a nail instead of a claw. The teeth and hands act together as the Aye - aye forages for the larvae insects, its main diet. It taps wood harboring insects and listens for them. If necessary, it uses the incisors to tear wood from the trees. The third digit is used to remove the insects from the holes in the wood and to dig the pulp out of certain fruits. Aye - ayes are critically endangered.

Danbentonia madagascariensis   Aye - Aye (E)
O   N.W.and E.Madagascar
          Z   W   P.W
                                                                P.Z

FAMILY: LORISIDAE, Lorises, Pottos, Galagos

The appearance of lorisids is less primitive than that of lemurids. The eyes face forward, the rostrum is short, the braincase is globular. The Potto is distinct in that some of its vertebrae project through the skin of the back of the neck. They are surrounded by long guard hairs. Lorisids inhabit forested areas of Africa south of the Sahara, India, Sri Lanka,and southeast Asia into the East Indies
Arboreal and nocturnal, lorises and pottos move in a methodical hand over hand fashion. They move quickly at times,but do not leap or jump. In contrast the galagos (often put in a separate family Galagidae), climb quite rapidly and jump from branch to branch. On the ground they hop or leap on their hind limbs. the diet is mainly carnivorous and insectivorous. The prey is usually captured by the hand after a stealthy approach.

Arctocebus calabarensis   Angwantibo (T*)
O   C. and W.Africa
          Z   W   P.W
                                                                P.Z
Loris tardigradus   Slender loris
O   S.Asia, Sri Lanka
          Z   W   P.W
                                                                P.Z

Nycticebus coucang	Slow loris		
O Bangladesh to Philippines			
Z	W		P.W
			P.Z
Nycticebus pygmaeus	Lesser slow loris (T*)		
O Viet Nam ,Laos			
Z	W		P.W
			PZ
Perodicticus potto	Potto		
O Ghana to Kenya			
Z	W		P.W
			P.Z
Euoticus elegantulus	Needle clawed bushbaby		
O Nigeria to Congo			
Z	W		P.W
			P.Z
Euoticus inustus	East needle clawed bushbaby		
O Zaire,Uganda			
Z	W		P.W
			P.Z
Galago alleni	Allen's bushbaby		
O Nigeria to Zaire			
Z	W		P.W
			P.Z
Galago crassicaudatus	Thick tailed bushbaby		
O E. and S. Africa			
Z	W		P.W
			P.Z
Galago demidovll	Demidoff's dwarf bushbaby		
O Africa			
Z	W		P.W
			P.Z
Galago senegalensis	Lesser bushbaby		
O Africa south of the Sahara			
Z	W		P.W
			P.Z

SUBORDER: TARSIOIDEA.

FAMILY: TARSIIDAE, Tarsiers

Roughly the size of a rat, the tarsier has a large head, a short neck, huge eyes facing forward with enormous orbits, long limbs, and a long tail. The head rotates through 180 degree, allowing an extremely wide field of vision.

The name tarsier (genus Tarsius), refers to elongation of 2 tarsal bones (ankle region) which make the animal a highly specialized jumper . The metatarsals (foot bones), Usually elongate in jumping animals, are not elongate, thus allowing the arboreal tarsier to grasp. Tarsiers leap from branch to branch with great precision. They can walk and climb quadrupedally, slide down trees, and also hop on the hind limbs on the ground. Tarsiers occur in jungles and secondary growth of southern Sumatra, some East Indian Islands, and some of the Philippine Islands.

They are nocturnal, the eyes well adapted for night vision. They feed primarily on insects, pouncing upon them and capturing them with the hands. They live in pairs and although usually silent, they are capable of making a variety of high pitched sounds.

Tarsius bancanus			Western tarsier (I)
O	Borneo, Sumatra,Java		
	Z	W	P.W
			P.Z
Tarsius spectrum			Celebesian tarsier (I)
O	Celebes		
	Z	W	P.W
			P.Z
Tarsius syrichta			Philippine tarsier
O	Philippines		
	Z	W	P.W
			P.Z

SUBORDER: ANTHROPOIDEA

FAMILY: CALLITRICHIDAE, Marmosets, Tamarins

The squirrel- sized marmosets and tamarins are slender - bodied with a long tail, elongated hands and feet, a sparsely haired face and hairy adornments about the head. Often they are conspicuously colored.

They are arboreal, inhabiting tropical forests of Panama , northern South America, and Brazil. They climb about in squirrel- like fashion,using their clawed digits. The thumb is not opposable; the big toe is opposable, and bears a nail.

Marmosets are omnivorous. The diet consists mostly of fruit and insects, though lizards, small buds, and eggs may be eaten by some species. They live in family groups and are quite social and extremely vocal. Typically twins are born; the father plays a major role in rearing and protecting the young.

The teeth differ in marmosets and tamarins. Tamarins have lower incisors of a normal length - - the canines are larger. Marmosets have elongated lower incisors so that the incisors and canines are approximately the same size.

Callimico goeldii			Goeldi's marmoset (R)
O	Amazonian rain forest		
	Z	W	P.W
			P.Z
Callithrix argentata			Silvery marmoset (V)
O	Brazil, Bolivia		
	Z	W	P.W
			P.Z
Callithrix aurita			White eared marmoset (E)
O	Rio de Janeiro ,Brazil		
	Z	W	P.W
			P.Z
Callithrix chrysoleuca			Yellow legged marmoset
O	Rio Madeira, Brazil		
	Z	W	P.W
			P.Z
Callithrix flaviceps			Buff headed marmoset (E)
O	Esp. Santo,Brazil		
	Z	W	P.W
			P.Z

Callithrix geoffroyi		White fronted marmoset
O Minas Gereus, Brazil		
Z	W	P.W
		P.Z

Callithrix humeralifer		Santarem marmoset (V)
O Rio Tapajos, Brazil		
Z	W	P.W
		P.Z

Callithrix jacchus		Common marmoset (V)
O Para to Bahia, Brazil		
Z	W	P.W
		P.Z

Callithrix penicillata		Black eared marmoset
O Brazil, Paraguay		
Z	W	P.W
		P.Z

Callithrix pygmaea		Pygmy marmoset
O Brazil, Peru		
Z	W	P.W
		P.Z

Leontopithecus rosalia		Golden marmoset (E)
O S.E.Brazil		
Z	W	P.W
		P.Z

Saguinus bicolor		Pied tamarin (I)
O N.E.Amazonas, Brazil		
Z	W	P.W
		P.Z

Saguinus devillei		Deville's tamarin
O Amazonas		
Z	W	P.W
		P.Z

Saguinus fuscicollis		Saddleback tamarin
O W. C.Amazonas, Brazil, Peru		
Z	W	P.W
		P.Z

Saguinus fuscus		Brown tamarin
O Amazonas		
Z	W	P.W
		P.Z

Saguinus geoffroyi		Geoffroy's tamarin
O Amazonas		
Z	W	P.W
		P.Z

Saguinus graellsi		Rio Napo tamarin
O N. Peru		
Z	W	P.W
		P.Z

Saguinus illigeri		Red mantled tamarin
O Colombia		
Z	W	P.W
		P.Z

Saguinus imperator   Emperor tamarin (I)
O   Brazil,Peru,Bolivia
    Z   W   P.W
                                                P.Z

Saguinus inustus   Mottle faced tamarin
O   Colombia,Brazil
    Z   W   P.W
                                                P.Z

Saguinus labiatus   White lipped tamarin
O   Amazonian region of Brazil
    Z   W   P.W
                                                P.Z

Saguinus lagonotus   Golden mantled tamarin
O   Amazonas
    Z   W   P.W
                                                P.Z

Saguinus leucopus   White footed tamarin (V)
O   Colombia
    Z   W   P.W
                                                P.Z

Saguinus martinsi   Martin's tamarin
O   Brazil
    Z   W   P.W
                                                P.Z

Saguinus melanoleucus   White tamarin
O   S.W.Amazona, Brazil
    Z   W   P.W
                                                P.Z

Saguinus midas   Red handed tamarin
O   Guinas,Brazil
    Z   W   P.W
                                                P.Z

Saguinus mystax   Moustached tamarin
O   Peru, Brazil
    Z   W   P.W
                                                P.Z

Saguinus nigricollis   Black and red tamarin
O   Amazonian region
    Z   W   P.W
                                                P.Z

Saguinus oedipus   Cotton headed tamarin (E)
O   Panama to Colombia
    Z   W   P.W
                                                P.Z

Saguinus pileatus   Red capped tamarin
O   W. Central Amazonas, Brazil
    Z   W   P.W
                                                P.Z

Saguinus pluto   Lonnberg's tamarin
O   Cen. Amazonas, Brazil
    Z   W   P.W
                                                P.Z

Saguinus tamarin   Negro tamarin
O   N.E. Para, Brazil
    Z   W   P.W
                                                P.Z

Saguinus weddelli   Weddell's tamarin
O   N.Bolivia, W.Brazil
    Z   W   P.W
                                                P.Z

FAMILY : CEBIDAE, New World monkeys

Usually small and slim, cebids are arboreal, mostly diurnal monkeys, inhabiting the tropical forests from southern Mexico to southern Brazil. They climb by grasping, using hands, feet and in some genera the prehensile tail. The thumb is not truly opposable, but is functionally so. In 2 genera, however , it is vestigial. All digits bear nails. The nostrils are separated by a broad nasal septum and open to the side, a condition known as platyrrhine. The diet of most cebids consistd mainly of fruit supplemented with plant and animal matter. One species, the night monkey or douroucouli, is carnivorous and insectivorous, possibly feeding on bats. Cebids are quite active; most move quadrupedally through the trees with great speed. Some,such as the spider monkey,also swing rapidly below the branches using their arms and prehensile tail.Most are quite vocal, especially the howler which emits loud roaring sounds which can be heard at great distances. These vocalizations help keep the troop together and keep territories separate. Most cebids are social, living in groups. Not much is known of their behavior in the wild ,due to the inaccessibility of the habitat.

Aotus trivirgatus   Douroucouli
O   Panama to Argentina
    Z   W   P.W
                                                P.Z

Callicebus moloch   Dusky titi
O   N.South America
    Z   W   P.W
                                                P.Z

Callicebus personatus   Masked titi (V)
O   S.E.Brazil
    Z   W   P.W
                                                P.Z

Callicebus torquatus   Collared titi
O   N.South America
    Z   W   P.W
                                                P.Z

Cacajao calvus   Bald uakari (V)
O   Brazil, Peru, Colombia
    Z   W   P.W
                                                P.Z

Cacajao melanocephalus   Black headed uakari (V)
O   N.E.Amazonas,Colombia,Venezuela
    Z   W   P.W
                                                P.Z

Cacajao rubicundus   Red uakari (E*)
O   Brazil, Peru
    Z   W   P.W
                                                  P.Z

Chiropotes albinasus   White nosed saki (V)
O   Brazil
    Z   W   P.W
                                                  P.Z

Chiropotes satanas   Black saki (E)
O   N.South America
    Z   W   P.W
                                                  P.Z

Pithecia monachus   Monk saki
O   N. South America
    Z   W   P.W
                                                  P.Z

Pithecia pithecia   Pale headed saki
O   Venezuela, Colombia, Brazil
    Z   W   P.W
                                                  P.Z

Alouatta belzebul   Red handed howler
O   Brazil
    Z   W   P.W
                                                  P.Z

Alouatta caraya   Black howler
O   Bolivia, Brazil, Paraguay, Argentina
    Z   W   P.W
                                                  P.Z

Alouatta fusca   Brown howler (I)
O   Bolivia, Brazil, Argentina
    Z   W   P.W
                                                  P.Z

Alouatta palliata   Howler monkey
O   Mexico to Ecuador
    Z   W   P.W
                                                  P.Z

Alouatta seniculus   Red howler
O   N.South America
    Z   W   P.W
                                                  P.Z

Alouatta villosa   Mantled howler (I)
O   S.E.Mexico to Ecuador
    Z   W   P.W
                                                  P.Z

Cebus albifrons   White fronted capuchin
O   N.South America
    Z   W   P.W
                                                  P.Z

Cebus apella   Tufted capuchin
O   N.South America
    Z   W   P.W
                                                  P.Z

Cebus capucinus			White throated capuchin
O Honduras to Ecuador			
Z	W		P.W
			P.Z
Cebus nigrivittatus			Weeper capuchin
O N.South America			
Z	W		P.W
			P.Z
Saimiri oerstedii			Squirrel monkey (E*)
O Costa Rica, Panama			
Z	W		P.W
			P.Z
Saimiri sciureus			Common squirrel monkey
O N.South America			
Z	W		P.W
			P.Z
Ateles belzebuth			Long haired spider monkey (V)
O N.South America			
Z	W		P.W
			P.Z
Ateles fusciceps			Brown headed spider monkey (I)
O Panama, Colombia, Ecuador			
Z	W		P.W
			P.Z
Ateles geoffroyi			Black handed spider monkey (V)
O Mexico to Colombia			
Z	W		P.W
			P.Z
Ateles paniscus			Black spider monkey (V)
O N.South America			
Z	W		P.W
			P.Z
Brachyteles arachnoides			Woolly spider monkey (E)
O Brazil			
Z	W		P.W
			P.Z
Lagothrix flavicauda			Yellow tailed woolly monkey (E)
O Peru			
Z	W		P.W
			P.Z
Lagothrix lagothricha			Woolly monkey (V)
O N.South America			
Z	W		P.W
			P.Z

FAMILY: CERCOPITHECIDAE, Old World monkeys

This large, diverse family lives mainly in the tropical forests and savannahs of Africa and Asia, although some members inhabit the high forests of Tibet and others live in northern Japan where snow falls. The size of these monkeys ranges greatly from the talapoin monkey (body size 14", less than 2 lbs.) to the large mandrills (about 100 lbs.)

In contrast with the New World monkeys, the nostrils of most Old World monkeys are close together and open downward, a condition referred to as catarrhine. The tail may be vestigial, short, or long, but is never prehensile. In the rump area, Old World monkeys have ischial callosities, callous pads with no sensitivity, which allow the animals to sit on branches or rocks comfortably, for long periods of time. All the digits bear nails and both the thumb and great toe are opposable, with the exception of the colobus monkey which has a vestigial thumb. The hands are used both in grasping and food handling. The senses of sight, smell and hearing are acute. There is much vocalization and use of facial expressions.

The family Cercopithecidae is sometimes divided into two sub- families. Members of the subfamily Colobinae (colobus, langurs, leaf monkeys), are primarily leaf eaters and the stomach is sacculated to accomodate this diet .All other Old World monkeys belong to sub- family Cercopithecinae. They have cheek pouches and most are omnivorous. Old World monkeys occupy all levels of the trees of the tropical forests and some have become well adapted to terrestrial life both in the forests and on the open savannahs. Most ground dwelling monkeys live in groups with complicated social behavior and a well-defined dominance hierarchy.

This adaptation, plus the great size of the male with his enlarged canines, has enabled these monkeys to successfully occupy this ecological niche and survive when threatened by predators

NOTE: Although some New World and Old World monkeys are quite similar in appearance, they evolved these adaptations separately, in different geographical locations from a prosimian ancestor. In similar habitats, similar favorable adaptations evolved, a process known as parallel evolution.

Cercocebus albigena   Gray cheeked mangabey
O   Cameroon to Kenya
    Z   W   P.W
                                                 P.Z

Cercocebus aterrimus   Black mangabey
O   Zaire, Angola
    Z   W   P.W
                                                 P.Z

Cercocebus galeritus   Tana river mangabey (E)
O   Guinea to Kenya
    Z   W   P.W
                                                 P.Z

Cercocebus torquatus   Sooty mangabey (E*)
O   Senegal to Zaire
    Z   W   P.W
                                                 P.Z

Cercopithecus aethiops   Savanna monkey
O   Ethiopia,Sudan
    Z   W   P.W
                                                 P.Z

Cercopithecus ascanius   Red tail guenon
O   Cameroon to Zambia
    Z   W   P.W
                                                 P.Z

Cercopithecus campbelli   Campbell's guenon
O   W.Africa
    Z   W   P.W
                                                 P.Z

Cercopithecus cephus			Moustached monkey
O W. Africa			
Z	W		P.W
			P.Z
Cercopithecus denti			Dents monkey
O Zaire, Uganda			
Z	W		P.W
			P.Z
Cercopithecus diana			Diana monkey (E*)
O Sierra Leone to Ghana			
Z	W		P.W
			P.Z
Cercopithecus dryas			Congo Diana monkey
O Zaire			
Z	W		P.W
			P.Z
Cercopithecus erythrogaster			Red bellied guenon (E*)
O Nigeria			
Z	W		P.W
			P.Z
Cercopithecus hamlyni			Owl faced monkey
O E. Congo basin			
Z	W		P.W
			P.Z
Cercopithecus lhoesti			Lhoest's monkey (E*)
O W.and E. Africa			
Z	W		P.W
			P.Z
Cercopithecus mitis			Diadem monkey
O Ethiopia to S. Africa			
Z	W		P.W
			P.Z
Cercopithecus mona			Mona monkey
O Ghana to Cameroon			
Z	W		P.W
			P.Z
Cercopithecus neglectus			De Brazzas monkey
O Cameroon to Zaire			
Z	W		P.W
			P.Z
Cercopithecus nictitans			White nosed monkey
O Guinea to Zaire			
Z	W		P.W
			P.Z
Cercopithecus nigroviridis			Allen's monkey
O Zaire			
Z	W		P.W
			P.Z
Cercopithecus patas			Patas monkey
O Senegal to Tanzania			
Z	W		P.W
			P.Z

Cercopithecus petaurista		Lesser spot nosed guenon	
O W. Africa			
Z	W		P.W
			P.Z
Cercopithecus pogonias		Crowned guenon	
O Guinea to Zaire			
Z	W		P.W
			P.Z
Cercopithecus talapoin		Talapoin monkey	
O C.and W. Africa			
Z	W		P.W
			P.Z
Cercopithecus wolfi		Wolf's monkey	
O Zaire			
Z	W		P.W
			P.Z
Macaca arctoides		Stump tailed macaque	
O S.E.Asia			
Z	W		P.W
			P.Z
Macaca assamensis		Assamese macaque (T*)	
O Assam to Burma, Himalayas			
Z	W		P.W
			P.Z
Macaca cyclopis		Taiwan macaque (T*)	
O Taiwan			
Z	W		P.W
			P.Z
Macaca fascicularis		Crab eating macaque	
O S.E.Asia			
Z	W		P.W
			P.Z
Macaca fuscata		Japanese macaque (T*)	
O Japan			
Z	W		P.W
			P.Z
Macaca mulatta		Rhesus macaque	
O S.Asia			
Z	W		P.W
			P.Z
Macaca nemestrina		Peg tailed macaque	
O S.E.Asia			
Z	W		P.W
			P.Z
Macaca nigra		Black ape	
O Celebes			
Z	W		P.W
			P.Z
Macaca radiata		Bonnet monkey	
O India			
Z	W		P.W
			P.Z

Macaca silenus		Lion tailed macaque (E)	
O India			
Z	W		P.W
			P.Z
Macaca sinica		Toque monkey (T*)	
O Sri Lanka			
Z	W		P.W
			P.Z
Macaca sylvanus		Barbary ape	
O Morocco, Algeria, Gilbraltar			
Z	W		P.W
			P.Z
Macaca thibetana		Stump tailed macaque	
O China. Szechuan			
Z	W		P.W
			P.Z
Papio cynocephalus		Yellow baboon	
O W.E. and S. Africa			
Z	W		P.W
			P.Z
Papio hamadryas		Hamadryas baboon	
O Ethiopia, Sudan, Somalia, Arabian peninsula			
Z	W		P.W
			P.Z
Papio leucophaeus		Drill (E)	
O W.Africa			
Z	W		P.W
			P.Z
Papio sphinx		Mandrill (E*)	
O W.Africa			
Z	W		P.W
			P.Z
Theropithecus gelada		Gelada baboon (E)	
O Ethiopia			
Z	W		P.W
			P.Z
Colobus badius		Red colobus	
O Senegal to Kenya			
Z	W		P.W
			P.Z
Colobus guereza		Guereza colobus	
O Nigeria to Tanzania			
Z	W		P.W
			P.Z
Colobus kirkii		Kirk's colobus (R)	
O Zanzibar Isl.			
Z	W		P.W
			P.Z
Colobus polykomos		Black and white colobus	
O E. to W. Africa			
Z	W		P.W
			P.Z

Nasalis larvatus			Proboscis monkey (V)
O	Borneo		
	Z	W	P.W
			P.Z

Presbytis aygula			Sunda Isl. leaf monkey
O	Java, Borneo, Sumatra		
	Z	W	P.W
			P.Z

Presbytis cristata			Silvered leaf monkey
O	S.Asia, Java, Borneo, Sumatra		
	Z	W	P.W
			P.Z

Presbytis entellus			Entellus langur (E*)
O	Tibet, Nepal, India		
	Z	W	P.W
			P.Z

Presbytis francoisi			Francois monkey (E*)
O	S.China, Viet Nam, Laos		
	Z	W	P.W
			P.Z

Presbytis frontata			White fronted leaf monkey
O	Borneo		
	Z	W	P.W
			P.Z

Presbytis geei			White fronted leaf monkey (R)
O	S.E.Asia		
	Z	W	P.W
			P..Z

Presbytis hosei			Leaf monkey
O	Borneo		
	Z	W	P.W
			P.Z

Presbytis johnii			John's langur
O	S.W.Asia		
	Z	W	P.W
			P.Z

Presbytis melalophos			Banded leaf monkey
O	Malay peninsula, Sumatra, Borneo		
	Z	W	P.W
			P.Z

Presbytis obscura			Dusky leaf monkey
O	Malay peninsula		
	Z	W	P.W
			P.Z

Presbytis phayrei			Phayre's leaf monkey
O	S.E.Asia		
	Z	W	P.W
			P.Z

Presbytis pileata		Capped langur (E*)		
O S.Asia				
Z	W			P.W
				P.Z
Presbytis potenziani		Mentawai leaf monkey (I)		
O Sumatra				
Z	W			P.W
				P.Z
Presbytis rubicundus		Maroon leaf monkey		
O Borneo				
Z	W			P.W
				P.Z
Presbytis senex		Purple faced langur (T*)		
O Sri Lanka				
Z	W			P.W
				P.Z
Presbytis thomasi		Silvered leaf monkey		
O Sumatra				
Z	W			P.W
				P.Z
Procolobus verus		Green colobus		
O W. Africa				
Z	W			P.W
				P.Z
Pygathrix nemaeus		Douc langur (E)		
O Laos, Viet Nam				
Z	W			P.W
				P.Z
Rhinopithecus avunculus		Tonkin snub nosed monkey (T*)		
O N. Viet Nam				
Z	W			P.W
				P.Z
Rhinopithecus brelichi		Golden monkey		
O China, N. Kweichow				
Z	W			P.W
				P.Z
Rhinopithecus roxellanae		Snub nosed monkey		
O China				
Z	W			P.W
				P.Z
Simias concolor		Pig tailed langur		
O Sumatra				
Z	W			P.W
				P.Z

FAMILY : PONGIDAE, Gibbons, Siamangs, Great Apes

Gibbons or " lesser Apes", are the most agile of the primates and the most spectacular brachiators of the mammals. Their extremely long arms are held up in the air for balance when on the ground. Most of their time is spent in the trees. The hands are adapted for brachiation - - digits 2 - 5 are long and used in a hook - like fashion during brachiation .There is little sexual dimorphism ,males and females are about the same size.Gibbons live in small family groups in very small territories in the tropical forests of Southeast Asia ,

Sumatra, Java, Borneo, and the Mentawi Islands. Their territory is vigorously defended with vocalizations. The largest of the gibbons, the siamang, has a naked throat sac which can be inflated to the size of its head. It is used to give loud resonating hoots which can be heard for great distances.

Chimpanzees and gorillas of Africa and orangutans of Borneo and Sumatra are known as the great apes. All are declining. Orangs and mountain gorillas are endangered. Chimpanzees and lowland gorillas are vulnerable. They are noted for their large size, robust bodies, and powerful arms which are longer than the legs . They lack a tail. The hair is short and usually sparse, the face is nearly naked. The trunk is short and wide, in comparison with the longer, narrow trunk of monkeys. The hands are large with digits 2 - 5 long and the thumb relatively short, all with nails. The hand can be used for grasping.

Gorillas are primarily terrestrial, orangutans arboreal, and chimpanzees spend time both in the trees and on the ground. When on the ground ,the normal mode of progression of the chimp and gorilla is quadrupedal. The hands are not flattened palm down, but the fingers are curled under and the animals are said to "Knucklewalk". Both can walk in an upright position which frees the hands for feeding of for carrying objects. Due to their size, orangutans move rather slowly and deliberately in the trees, a contrast to the small and agile gibbons.

The social structure and diet of the three great apes is quite different. Chimpanzees live in highly complicated social units with much interaction and vocalization between individuals. The groups called bands, are fluid; group composition changes frequently. Some dominance is shown among males. Some females also assume a position in the dominance hierarchy. Omnivorous, chimps are opportunists, feeding on fruits, vegetable matter, insects and whatever is available. Some are reported to eat meat (primarily monkeys and baboons) which they hunt in organized fashion.

Gorillas live in small groups led by a dominant male- usually a "silverback". They are very quiet and show little interaction with each other. They spend most of the day foraging and are strictly vegetarian in the wild. They are the largest of the primates, males in the wild reaching 400 lbs.

Orangutans often live alone or in pairs or in small family groups, these small numbers may be due to the low numbers of orangs and their dispersal throughout the forests. They make little noise and feed mainly on the fruit of the Durian Tree. All of the apes construct "nests" at night. This consists of putting together branches or leaves for sleeping. Most lack ischial callosities. They are all nomadic, sleeping in a different place each night. They usually move within a set range. The only lasting bond between individuals is that between the mother and her young. Infants stay with their mothers for a long period of time. In chimps the bond may last 6 years of more.

Hylobates agilis   Dark haired gibbon (E*)
O   Malaysia, Sumatra
        Z   W   P.W
                                                              P.Z

Hylobates concolor   Black gibbon (I)
O   China, Laos, Viet Nam, Cambodia
        Z   W   P.W
                                                              P.Z

Hylobates hoolock   Hoolock Gibbon
O   China, Burma
        Z   W   P.W
                                                              P.Z

Hylobates klossi   Dwarf gibbon (V)
O   Mentawi Isl. (off Sumatra)
    Z   W   P.W
                              P.Z

Hylobates lar   White handed gibbon (E*)
O   S.E.Asia
    Z   W   P.W
                              P.Z

Hylobates moloch   Gray gibbon (E)
O   Java, Borneo
    Z   W   P.W

                              P.Z

Hylobates pileatus   Pileated gibbon (E)
O   Thailand, Cambodia
    Z   W   P.W
                              P.Z

Hylobates syndactylus   Siamang (E*)
O   Malaysia
    Z   W   P.W
                              P.Z

Gorilla gorilla   Gorilla (V)
O   Equatorial Africa
    Z   W   P.W

                              P.Z

Pan paniscus   Pygmy chimpanzee (V)
O   Zaire
    Z   W   P.W
                              P.Z

Pan troglodytes   Chimpanzee (E)
O   W.to E. Africa
    Z   W   P.W
                              P.Z

Pongo pygmaeus   Orangutan (E)
O   Borneo, Sumatra
    Z   W   P.W
                              P.Z

Man is the culmination of many primate trends. He has a greatly enlarged brain, a flattened face, an opposable thumb with both a power and a precision grip. He is the only completely bipedal terrestrial primate. Using his great intelligence, he has adapted the world about him to fit his needs and has spread out to occupy all parts of the earth. He is the most efficient predator ever to evolve; yet in man's intelligence lies the hope for a future for all the other animals on earth.

Homo sapiens   Man
    World wide

TOTAL PRIMATES: O	Z	W	P.W.	P.Z.

ORDER: PRIMATES

ORDER: PRIMATES

ORDER: PRIMATES

ORDER: PRIMATES

ORDER: PRIMATES

ORDER: PRIMATES

ORDER: EDENTATA

Formerly widespread and abundant, the surviving members of the order anteaters, sloths, and armadillos, are found only in the New World.

Although quite different both in appearance and habits, they share a series of distinctive structural features. Edentata means "without teeth'. However, only the anteaters do not have teeth. Sloths and armadillos have reduced and simplified dentition.

The brain is small, the braincase usually long and cylindrical. The limbs are usually specialized for digging or climbing, the digits ending in long, strong claws.

Some fossil edentates were of an unusually large size. Megatherium, a massive ground sloth larger than an elephant, lived in South America during the Oligocene.

An armadillo to the Pleistocene in South America was the size of a rhinoceros.

FAMILY: MYRMECOPHAGIDAE, Anteaters

Anteaters are highly specialized for feeding on ants and termites. The jaw bones are long and delicate: there are no teeth; the tongue is long and protrusile, covered with a sticky saliva. The forelimbs are powerful, the enlarged third digit bearing a stout recurved claw. The terrestrial giant anteater has 3 large claws on the forefoot and walks on its knuckles with the digits partly flexed. Arboreal anteaters have a prehensile tail and walk on the sides of the forefeet, the claws turned inward.

Anteaters forage for insects in the tropical forests of central and south America. Ant and termite nests are torn apart and insects captured on their long, sticky tongue. The insects are swallowed whole and ground up in the stomach.

To defend themselves against predators, giant anteaters stand on the hind limbs and use the tail to form a tripod, slashing at an enemy with the claws of the forefoot

Anteaters bear one offspring which clings to the back of the giant anteater and to the tail of the silk anteater.

Cyclopes didactylus   Silky anteater
O   S. Mexico to Brazil and Bolivia
    Z   W   P.W
                                    P.Z

Myrmecophaga tridactyla   Giant anteater (V)
O   Belize to Argentina
    Z   W   P.W
                                    P.Z

Tamandua mexicana   Tamandua
O   Mexico to Venezuela, Peru
    Z   W   P.W
                                    P.Z

Tamandua tetradactyla   Lesser anteater
O   S.America
    Z   W   P.W
                                    P.Z

FAMILY: BRADYPODIDAE, Sloths

Sloths are entirely arboreal, feeding on leaves in the tropical forests of central and northern South America. In contrast to other edentates, the sloth's skull is short, the nose are flattened. Most mammals have 7 cervical vertebrae, but sloths have from 6 to 9 depending on the species. Algae grow on the surfaces of the hairs during the rainy season and give the fur a greenish tint

The externally visible digits do not exceed 3 in number and are bound together for nearly their entire length with skin and tissues. The fingers and toes have long curved, strong claws. Spending their lives in trees, sloths are nearly helpless on the ground. They can swim when the forest floor is flooded. They climb in an upright position embracing the branch or by hanging upside down, and moving along hand over hand. They spend much time upside down, sometimes sleeping in that position

Sloths are apparently very sensitive to cold and are heterothermic (have a varying body temperature).There are 2 genera of sloths - - the three toed and two toed (referring to the number of claws on the forefoot)

Two toed sloths survive in captivity but three toed sloths do not do well, probably due to their highly specialized diet - - leaves of one kind of tree.

130

Bradypus torquatus   Maned sloth (E)
O   Brazil
    Z   W   P.W
                                             P.Z

Bradypus tridactylus   Three toed sloth
O   Guianas, Brazil, Venezuela
    Z   W   P.W
                                             P.Z

Bradypus variegatus   Three toed sloth
O   Honduras to Argentina
    Z   W   P.W
                                             P.Z

Choloepus didactylus   Two toed sloth
O   N. of South America
    Z   W   P.W
                                             P.Z

Choloepus hoffmann   Hoffman's sloth
O   Nicaragua to Peru, Brazil
    Z   W   P.W
                                             P.Z

FAMILY: DASYPODIDAE, Armadillos

The distinguishing feature of the armadillo is the protective armor which consists of bony scutes covered by horny epidermis. Sparse hair usually occurs on the skin between the plates and on the limbs and underside. Individuals of some species can curl into a ball and cover their vulnerable underparts. The limbs are powerful and all 4 feet bear large, heavy claws for digging. Some are fossorial (burrowers).

The most successful of the edentates, they occupy a variety of habitats; tropical forests, savannahs, semi- deserts, temperate plains and forests from south-central United States to the southern end of Argentina.

Armadillos feed mostly on insects but also on invertebrates, small vertebrates and vegetable material. They travel singly, in pairs, or occasionally in small bands.

Cabassous centralis   Naked tailed armadillo
O   Honduras to Colombia
    Z   W   P.W
                                             P.Z

Cabassous chacoensis   Naked tailed armadillo
O   South America
    Z   W   P.W
                                             P.Z

Cabassous tatouay   Armadillo
O   Argentina,Brazil,Paraguay,Uruguay
    Z   W   P.W
                                             P.Z

Cabassous unicinctus   Broad banded armadillo
O   Brazil, Venezuela
    Z   W   P.W
                                             P.W

Chaetophractus nationi   Armadillo
O   Argentina, Bolivia, Chile, Peru
    Z   W   P.W
                                             P.Z

Chaetophractus vellerosus   Long haired armadillo
O   Argentina, Bolivia, Paraguay
     Z   W   P.W
                                                      P.Z

Chaetophractus villosus   Hairy armadillo
O   Argentina, Paraguay
     Z   W   P.W
                                                      P.Z

Chlamyphorus truncatus   Pink fairy armadillo (E*)
O   Argentina
     Z   W   P.W
                                                      P.Z

Chlamyphorus retusus   Pichiciegos
O   Argentina, Bolivia, Paraguay
     Z   W   P.W
                                                      P.Z

Dasypus hybridus   Mulita armadillo
O   Argentina, Paraguay, Uruguay
     Z   W   P.W
                                                      P.Z

Dasypus kappleri   Armadillo
O   N.South America
     Z   W   P.W
                                                      P.Z

Dasypus novemcinctus   Nine banded armadillo
O   S.E. United States to Argentina
     Z   W   P.W
                                                      P.Z

Dasypus pilosus   Armadillo
O   Peru
     Z   W   P.W
                                                      P.Z

Dasypus sabanicola   Armadillo
O   Venezuela, Colombia
     Z   W   P.W
                                                      P.Z

Dasypus septemcinctus   Seven banded armadillo
O   N. South America
     Z   W   P.W
                                                      P.Z

Euphractus sexcintus   Six banded armadillo
O   Argentina, Brazil
     Z   W   P.W
                                                      P.Z

Priodontes maximus   Giant armadillo (V)
O   South America
     Z   W   P.W
                                                      P.Z

Tolypeutes matacus   La Plata three banded armadillo
O   N.South America
     Z   W   P.W
                                                      P.Z

ARMADILLOS

Tolypeutes tricinctus   Three banded armadillo (I)
O   Brazil
     Z   W   P.W
                                               P.Z

Zaedyus pichiy   Pichi
O   Argentina
     Z   W   P.W
                                               P.Z

TOTAL EDENTATA: O   Z   W   P.W.   P.Z.

ORDER: EDENTATA

ORDER:PHOLIDOTA

Pangolins or Scaly Anteaters, are covered with horny scales which give them a reptilian appearance. The skull is conical; the tongue is extremely long, narrow and covered with sticky saliva, the jaw is toothless. They inhabit tropical and subtropical parts of the southern half of Africa and much of southeast Asia.Like anteaters, pangolins are insectivorous and were once considered members of the order Edentata due to the many common characteristics shared with anteaters. It is thought that these similarities are not due to a genetic relationship but to convergent evolution. A condition in which animals not closely related develop similarities due to modes of life.Some pangolins are terrestrial; others arboreal. When attacked, they can roll into a ball for protection. When on the ground they "knuckle -walk" as the giant anteaters. They can also stand on their hind limbs using the tail as a support

FAMILY: MANIDAE, Pangolins

Manis crassicaudata | Indian pangolin
O   Sri Lanka, India

Manis gigantea | Giant pangolin
O   Senegal to Uganda

Manis javanica | Malayan pangolin
O   S.Asia

Manis pentadactyla | Chinese pangolin
O   China, S,Asia

Manis temmincki | Cape pangolin (E*)
O   E.and S.Africa

Manis tetracactyla | Tree pangolin
O   Senegal to Uganda

Manis tricuspis | African tree pangolin
O   W.to E. Africa

TOTAL PHOLIDOTA: O   Z   W   P.W   P.Z

ORDER:LAGOMORPHA

Although not a diverse order, the lagomorphs (pikas, rabbits,and hares), have been very successful in terms of their geographic distribution.

They occur naturally on all large land masses except the Australian region and southern South America but recently have been introduced into these regions by man. They now occupy many types of terrestrial habitats from the arctic to the tropics.

Lagomorphs were once classified as rodents on the basis of superficial similarities, but fossil records indicate the two groups evolved from completely different ancestors.

Rodents have two upper incisors which are ever - growing. Lagomorphs have four upper incisors; two are small and peglike and lie directly behing the other two. There is a large diastema (gap) between the incisors and cheek teeth due to the absence of canines

The front feet have 5 digits, the hind feet 4 or 5 . The soles of the feet are covered with hair. Although the foot posture is plantigrade for slow movement, the running posture is digitigrade.

Lagomorphs are herbivorous, preferring grasses and plants but resorting to bark, shrubs and stems when food is scarce. The fecal material is of two types; moist "night" pellets which are expelled and eaten directly from the anus (Leporids) or from the floor of the den (pikas),and dry pellets which are not eaten. This allows most of the food to go through the digestive system twice and therefore more nutritional value is obtained..

Pikas are not as specialized for running as are rabbits and hares. They generally venture only short distances from shelter. They are quite small, usually 6 or 7 inches long, weighing only about 1/4 lb. ,with short rounded ears, short legs,and no visible tail.Pikas inhabit alpine and boreal areas from sea level in Alaska, to the treeless peaks of high mountains in the Rockies and Sierra cascade ranges in North America. Usually their home includes a fairly extensive rocky area with adjacent plant growth. They are diurnal and colonial. In Eurasia, they have a wider distribution and occupy a wider range of habitats. Of interest are the large "hay piles" built during the summer and used as a food source during the winter. This remarkable behavioral adaptation is correlated with the pikas' inability to hibernate.

Ochotona alpina   Altai pika
O   Altai to Japan

Ochotona collaris   Collared pika
O   Eastern Alaska, Northwestern Canada

Ochotona daurica   Daurian pika
O   Altai to Tibet to Iran

Ochotona hyperborea   Northern pika
O   Siberia to Japan

Ochotona koslowi   Koslow's pika
O   Northern Tibet

Ochotona ladacensis   Ladak pika
O   Southern Sinkiang, Kashmir

Ochotona pallasi   Pallas pika
O   Sinkian, Mongolia, Soviet Union

Ochotona princeps   American pika
O   S.W. Canada and western United States

Ochotona pusilla   Steppe pika
O   Soviet Union, S.Asia

Ochotona roylei   Royle's pika
O   Tibet, W.China. N.India

Ochotona rufescens   Afghan pika
O   Afghanistan, Turkmen,Iran

Ochotona rutila   Red pika
O   Kirghiz to Afghanistan

Ochotona thibetana   Mountain pika
O   Sikkim, China, Burma

Leporids are a very successful group occupying a variety of habitats over broad areas. They are more highly specialized for running than are the pikas - - the limbs, especially the hind ones, are elongate.The ears are long and tubular in shape; the tail short, the teeth specialized for a completely herbivorous diet. Within the family there are differences in running ability which correlate with habitat preference. Those that live in dense vegetation are not as specialized for running - - they scamper to safety and hide. Those that live in the more open areas rely on speed for protection. One of these is the jackrabbit whose hind limbs are greatly elongated allowing him to attain speeds of over 40 mph. Leporids of the genus Lepus and some other genera are called hares. They live in open areas. The young are ready to move about soon after birth . In contrast, Leporids called rabbits, occupy burrows called warrens and give birth to altricial young - - blind and naked- - in a nest of vegetation and fur. Common names often are erroneous as the "jackrabbit" is really a hare and Belgian hare is really a rabbit.

Lepus alleni   Antelope jackrabbit
O   Arizona., New Mexico, Mexico

Lepus americanus   Snowshoe
O   Alaska,Canada, N.United States

Lepus arcticus   Arctic hare
O   Tundra zone of the Arctic

Lepus brachyurus   Japanese hare
O   Japan, E.Asia

Lepus californicus   Black tailed jackrabbit
O   W.and C. United States,Mexico

Lepus callotis   White sided jackrabbit
O   S.W. New Mexico, Mexico

Lepus capensis   Cape hare
O   Palearctic region, Africa

Lepus crawshayi   Crawshay's hare
O   Africa

Lepus europaeus   European hare
O   Palearctic region

Lepus flavigularis   Tehuantepec jackrabbit
O   Oaxaca , Mexico

Lepus gaillardi
O   New Mexico, Mexico

Gaillards Jackrabbit

Lepus habessinicus
O   Somalia, Ethiopia

Abyssinian hare

Lepus insularis
O   Espiritu Santo Isl, (Baja California)

Black jackrabbit

Lepus mexicanus
O   Central Mexico

White sided jackrabbit

Lepus monticularis
O   South Africa

Bushman hare

Lepus nigricollis
O   Indian, Sri Lanka

Indian hare

Lepus oiostolus
O   Tibet, N.India

Woolly hare

Lepus othus
O   Alaska

Tundra hare

Lepus peguensis
O   Indochina

Burmese hare

Lepus saxatilis
O   South Africa

Scrub hare

Lepus siamensis
O   Thailand, Laos, Burma

Siamese hare

Lepus sinensis
O   Korea, China, Taiwan

Chinese hare

Lepus timidus
O   Europe (tundra and coniferous)

Blue hare

Lepus townsendi
O   S.Canada, Uniteds States

White tailed jackrabbit

Lepus whytei
O   Malawi

Whyte's hare

Lepus yarkandensis
O   Sinkiang

Yarkand hare

Nesolagus netscheri
O   Sumatra

Short eared rabbit (R)

Oryctolagus cuniculus   **Domestic rabbit**
O   original range was southwestern Europe, N.W.Africa
                                    P

Pentalagus furnessi   Ryukyu rabbit (E)
O   Ryukyus islands

Pronolagus crassicaudatus   Natal red hare
O   South Africa

Pronolagus randensis   Rand red hare
O   S.W.Africa

Pronalagus rupestris   Smith's red hare
O   Kenya to south Africa

Romerolagus diazi   Volcano rabbit
O   Mt. Popocatepetl, (Mexico)

Sylvilagus aquaticus   Swamp rabbit
O   South - Central United States

Sylvilagus audubonii   Desert cottontail
O   Western United States

Sylvilagus bachmani   Brush rabbit
O   Washington to Baja California

Sylvilagus brasiliensis   Forest rabbit
O   Mexico to Argentina

Sylvilagus cunicularius   Mexican cottontail
O   Mexico

Sylvilagus floridanus   Eastern cottontail
O   Eastern Canada to Venezuela

Sylvilagus graysoni   Tres Marias cottontail
O   Tres Marias Isl. (Mexico)

Sylvilagus idahoensis   Pygmy rabbit
O   Western United States

Sylvilagus insonus   Omilteme cottontail
O   Guerrero,(Mexico)

Sylvilagus mansuetus
O   San Jose Isl, (Baja California)

San Jose brush rabbit

Sylvilagus nuttalli
O   S.W Canada ,W. United States

Mountain cottontail

Sylvilagus palustris
O   Virginia to Florida

Marsh rabbit

Sylvilagus transitionalis
O   Maine to Alabama

New England cottontail

TOTAL LAGOMORPHA: O	Z	W	P.W	P.Z.

ORDER: LAGOMORPHA

ORDER: LAGOMORPHA

ORDER:RODENTIA

Rodents comprise the largest mammalian order and account for more species than all other mammals combined.

They are world wide in distribution and occupy a variety of niches in arboreal, fossorial, terrestrial, and aquatic habitats.

A wealth of structural diversity makes classification whithin the order difficult. This diversity is associated largely with different feeding habits and different styles of locomotion.

The most distinctive feature of rodents is the pair of ever-growing upper and lower incisors-an adaptation for their primarily herbivorous diet.

The incisors are covered with enamel on the front surface only- - their continuous use keeps them worm down and sharp. ·The cheek teeth, which are separated from the incisors by a wide diastema (gap), are complex and are used for grinding plant materials. In some species the cheek teeth are also evergrowing.

Some rodents have cheek pouches. They may be either internal or external and are used for food storage. External pouches are fur lined and may be turned inside out for cleaning. The skeleton in most forms, except for the head, is not highly specialized.

The order is often divided into 3 groups based on the shape of the jaw and its type of musculature.

Theses groups are:

SCIUROMORPHS : (squirrel- like)

MYOMORPHS : (mouse- like)

HYSTRICOMORPHS : (porcupine- like)

SCIUROMORPHS

FAMILY: APLODONTIDAE, Mountain beaver

The sewelled or mountain beaver of the Pacific Northwest is regarded as the most primitive living rodent.

Aplodontia rufa   Mountain beaver
O   British Columbia to N.W. California

FAMILY : SCIURIDAE, Squirrels, Chipmunks, Marmots(groundhogs), Prairie dogs

This is an extremely successful family. Sciurids have relatively unspecialized bodies; a long tail, no loss of digits, no reduction in freedom of the elbow, wrist or ankle joints, sharp claws. The profile of the head is usually arched. Cheek teeth are rooted. There are several semi- fossorial types and many arboreal species. One group of 13 genera known as flying squirrels, have a gliding surface formed by folds of skin between forefoot and hindfoot which unable them to glide between trees

Ammospermophilus harrisii   Yuma antelope squirrel
O   Sonora (Mexico), Arizona, New Mexico

Ammospermophilus insularis   Espiritu Island antelope squirrel
O   Espiritu Santo Isl. (Baja California)

Ammospermophilus interpres   Texas antelope squirrel
O   Texas, New Mexico to N.C.Mexico

Ammospermophilus leucurus   White tailed antelope squirrel
O   Oregon to Baja California

Ammospermophilus nelsoni   San Joaquin antelope squirrel
O   S.C. California

Atlantoxerus getulus   Barbary ground squirrel
O   North Africa

Callosciurus caniceps   Gray bellied squirrel
O   Burma to Malaysia

Callosciurus ferrugineus   Burma squirrel
O   Burma

Callosciurus finlaysoni   Finlayson's squirrel
O   Burma to Indochina

Callosciurus flavimanus   Belly banded squirrel
O   S.China to Indochina

145

Callosciurus inornatus   Tree squirrel
O   Viet Nam, Laos

Callosciurus melanogaster   Sumatra squirrel
O   Mentawi Isl.

Callosciurus nigrovittatus   Malayan squirrel
O   Thailand to Borneo

Callosciurus notatus   Spotted squirrel
O   S.E.Asia

Callosciurus prevosti   Prevost's squirrel
O   S.E.Asia

Callosciurus pygerythrus   Irawadoy squirrel
O   Nepal to Indochina

Callosciurus quinquestriatus   Anderson's squirrel
O   Yunnan,(China), Burma

Callosciurus sladeni   Burma squirrel
O   Burma

Cynomys gunnisoni   Gunnison's prairie dog
O   Utah, Colorado, Arizona, New Mexico

Cynomys leucurus   White tailed prairie dog
O   Montana, Wyoming, Utah, Colorado

Cynomys ludovicianus   Black tailed prairie dog
O   From Saskatchewan, Montana to Mexico

Cynomys mexicanus   Mexican prairie dog (E*)
O   Coahuila, (Mexico)

Cynomys parvidens   Utah prairie dog (V)
O   S.C.Utah

Dremomys everetti   Mountain ground squirrel
O   Borneo

Dremomys lokriah   Himalayan squirrel
O   Nepal to Burma

Dremomys pernyi   Perny's long nosed squirrel
O   China, Burma, Taiwan

Dremomys pyrrhomerus   Tree squirrel
O   China, N. Viet Nam

Dremomys rufigenis   Red cheeked ground squirrel
O   China, and S.E.Asia

Epixerus ebii   Ebien squirrel
O   Ghana to Sierra Leone

Epixerus wilsoni   African palm squirrel
O   Gabon

Eutamias alpinus   Alpine chipmunk
O   Sierra Nevada in E.C. California

Eutamias amoenus   Yellow pine chipmunk
O   S.W.Canada ,N.W.United States

Eutamias bulleri   Buller's chipmunk
O   Durango (Mexico)

Eutamias canipes   Gray footed chipmunk
O   New Mexico, Western Texas

Eutamias cinereicollis   Gray collared chipmunk
O   Arizona, New Mexico

Eutamias dorsalis   Cliff chipmunk
O   S.W.United States to Mexico

Eutamias merriami   Merriam's chipmunk
O   S.California, Baja California

Eutamias minimus   Least chipmunk
O   Canada, Western United States

Eutamias palmeri   Palmer's chipmunk
O   Nevada,(Charleston mountains)

Eutamias panamintinus   Panamint chipmunk
O   E.California, W. Nevada

Eutamias quadrimaculatus   Long eared chipmunk
O   E.C. California

Eutamias quadrivittatus   Colorado chipmunk
O   Utah, Colorado, Arizona,Washington,Idaho, Montana

147

Eutamias ruficaudus   Red tailed chipmunk
O Bristish Columbia,Alberta,Washington,Idaho,Montana

Eutamias sibericus   Siberian chipmunk
O N. Russia to C. China

Eutamias sonomae   Sonoma chipmunk
O N.W.California

Eutamias speciosus   Lodgepole chipmunk
O Mountains of E.and S. California

Eutamias townsendii   Townsend's chipmunk
O S.W.British Columbia to N. California

Eutamias umbrinus   Uinta chipmumk
O Nevada, Utah, Wyoming, Colorado

Exilisciurus concinnus   Pygmy squirrel
O Philippines

Exilisciurus exilis   Pygmy squirrel
O Borneo, Sumatra

Exilisciurus luncefordi   Pygmy squirrel
O Philippines (Mindanao)

Exilisciurus samaricus   Pygmy squirrel
O Philippines,(Samar)

Exilisciurus surrutilus   Pygmy squirrel
O Philippines,(Mindanao)

Exilisciurus whitheadi   Pygmy squirrel
O Borneo

Funambulus layardi   Striped squirrel
O S. India, Sri Lanka

Funambulus palmarum   Indian palm squirrel
O India, Sri Lanka

Funambulus pennati   North palm squirrel
O India, Pakistan,Nepal

Funambulus sublineatus   Dusky striped squirrel
O S. India, Sri Lanka

Funambulus tristriatus   Jungle striped squirrel
O   S.W.India

Funisciurus anerythrus   Redless squirrel
O   Gambia to Uganda

Funisciurus auriculatus   Matshcies squirrel
O   Nigeria to Zaire

Funisciurus bayoni   Bayon's squirrel
O   Angola, Zaire

Funisciurus carruthersi   Tanzania squirrel
O   Tanzania, Uganda

Funisciurus congicus   Congo striped squirrel
O   S.W.Africa

Funisciurus isabella   Lady Burton's squirrel
O   Cameroon, Central Africa Republic

Funisciurus lemniscatus   Western Africa striped squirrel
O   Cameroon, Gabon, Zaire, Congo

Funisciurus leugostigma   White spotted squirrel
O   Ivory Coast to Central Africa Republic

Funisciurus mandingo   Mandingo squirrel
O   Gambia to Nigeria

Funisciurus mystax   Rope squirrel
O   Congo, Zaire

Funisciurus pyrrnopus   Fire footed squirrel
O   Sierra Leone to Angola

Funisciurus raptorum   Striped squirrel
O   Nigeria

Funisciurus substriatus   Rope squirrel
O   Ivory Coast to Nigeria

Glyphotes canalvus   Sculptor squirrel
O   Borneo

Glyphotes simus   Bornean pygmy squirrel
O   Borneo

Guerlinguetus aestuans   Brazilian squirrel
O   Venezuela to Brazil,Argentina

Guerlinguetus cuscinus   Red squirrel
O   Peru

Guerlinguetus flammifer   Red squirrel
O   Venezuela

Guerlinguetus ignitus   Red squirrel
O   N.South America

Guerlinguetus igniventris   Red squirrel
O   Colombia, Bolivia, Brazil

Guerlinguetus iquiriensis   Red squirrel
O   Brazil

Guerlinguetus langsdorffi   Red squirrel
O   Brazil, Bolivia

Heliosciurus gambianus   Sun squirrel
O   Senegal to Sudan to Mozambique

Heliosciurus lucifer   Black and red squirrel
O   Tanzania, Malawi

Heliosciurus mutabilis   Sun squirrel
O   Tanzania

Heliosciurus punctatus   Sun squirrel
O   Mozambique

Heliosciurus rhodesiae   Sun squirrel
O   N.Zimbabwe

Heliosciurus rufobrachium   Red legged sun squirrel
O   Senegal to Sudan South to Mozambique

Heliosciurus undulatus   Sun squirrel
O   Tanzania

Hyosciurus heinrichi   Long snouted squirrel
O   Mountains of Celebes

Lariscus hosei   Borneo black striped squirrel (R)
O   Borneo

Lariscus insignis   Three striped ground squirrel
O   Malaysia to Borneo

Lariscus niobe   Striped squirrel
O   W.Sumatra

Lariscus obscurus   Striped squirrel
O   W.Sumatra, (Pagi Isl.)

Marmota bobak   Bobak marmot
O   Poland to Mongolia and China

Marmota caligata   Hoary marmot
O   Alaska to Washington and Montana

          Z   W   PW
                                                    P.Z
Marmota caudata   Long tailed marmot
O   Mountains of C. Asia

Marmota flaviventris   Yellow bellied marmot
O   S.W. Canada, Western Unites States
          Z   W   P.W
                                                    P.Z
Marmota marmota   Alpine marmot
O   Alpes to E.Siberia

Marmota monax   Woodchuck
O   Alaska, Canada, Eastern United States
          Z   W   P.W
                                                    P.Z
Marmota olympus   Olympic Mt. marmot
O   Olympic peninsula of Washington
          Z   W   P.W
                                                    P.Z
Marmota vancouverensis   Vancouver marmot (E*)
O   Vancouver Isl.

Menetes berdmorie   Berdmore's squirrel
O   S.E.Asia

Microsciurus alfari   Neotropical dwarf squirrel
O   Nicaragua to Panama

Microsciurus boquetensis   Pygmy squirrel
O   Chiriqui, Panama

Microsciurus flaviventer   Pygmy squirrel
O   Colombia, Ecuador

Microsciurus isthmius   Pygmy squirrel
O   S.Panama

Microsciurus medellineusis   Pygmy squirrel
O   Colombia

Microsciurus pucherani   Pygmy squirrel
O   Colombia

Myosciourus pumilio   African pygmy squirrel
O   Gabon, Cameroon

Nannosciurus luncefordi   Black eared squirrel
O   Philippines (Mindanao)

Nannosciurus melanotis   Black eared squirrel
O   Borneo

Nannosciurus samaricus   Samar squirrel
O   Philippines,(Samar)

Nannosciurus surrutilus   Mindanao squirrel
O   Philippines

Paraxerus alexandri   Striped squirrel
O   Zaire, Uganda

Paraxerus antoniae   African bush squirrel
O   Congo

Paraxerus boehmi   Boehmi's striped squirrel
O   Sudan to Zambia

Paraxerus byatti   Byatti squirrel
O   Tanzania

Paraxerus cepapi   Bush squirrel
O   S, Africa to Ethiopia

Paraxerus emini   Striped squirrel
O   Sudan to Congo

Paraxerus flavivittis   Eastern striped squirrel
O   Mozambique to Kenya

Paraxerus ochraceus   Bush squirrel
O   Sudan, Kenya, Tanzania

Paraxerus palliatus   South African red squirrel
O   Somalia to South Africa

Paraxerus sponsus   Red squirrel
O   Zululand (S.Africa)

Paraxerus vincenti   Vincent's squirrel
O   Mozambique

Paraxerus vulcanorum   Sriped squirrel
O   Congo, Zaire

Prosciurillus leucomus   Dwarf squirrel
O   Celebes,(Sanghir Isl)

Prosciurillus murinus   Celebes dwarf squirrel
O   Celebes

Prosciurillus obscurus
O   Celebes

Dwarf squirrel

Prosciurillus stangeri
O   Africa

African giant squirrel

Ratufa affinis
O   Malaya, Sumatra

Common giant squirrel

Ratufa bicolor
O   S.E.Asia to Nepal

Black giant squirrel

Ratufa indica
O   India

Malabar squirrel

Ratufa macroura
O   India, Sri Lanka

Grizzled indian squirrel

Rheithrosciurus macrotis
O   Borneo

Groove toothed squirrel

Rubrisciurus rubriventer
O   Celebes

Squirrel

Sciurillus pusillus
O   Guianas, Brazil, Peru

Neotropical pygmy squirrel

Sciurotamias davidianus
O   E.China

Rock squirrel

Sciurotamias forresti
O   China, (Yunnan)

Yunnan rock squirrel

Sciurus aberti
O   S.W.United States
     Z   W   P.

Tassel eared squirrel

Sciurus alleni
O   Mexico, (Nuevo Leon)

Allen's squirrel

Sciurus anomalus
O   Syria to S.W.Russia

Persian squirrel

Sciurus apache
O   Arizona, (Chiricahau Mts.)
     Z.   W   P

Apache fox squirrel

Sciurus arizonensis
O   Arizona, New Mexico, N.E.Mexico
     Z   W   P

Arizona gray squirrel

Sciurus aureogaster
O   Mexico, Guatemala

Red bellied squirrel

153

Sciurus carolinensis   Eastern gray squirrel
O   S.E.Canada. Eastern United States
     Z   W   P

Sciurus chiricahuae   Chiricahua squirrel
O   S.E.Arizona
     Z   W   P

Sciurus colliaei   Collie's squirrel
O   W. Mexico

Sciurus deppei   Deppe's squirrel
O   Mexico to Costa Rica

Sciurus griseoflavus   Guatemalan gray squirrel
O   Mexico,(Oaxaca) Guatemala

Sciurus griseus   Western gray squirrel
O   Washington to California
     Z   W   P

Sciurus kaibabensis   Kaibab squirrel
O   Arizona
     Z   W   P

Sciurus nayaritensis   Nayarit squirrel
O   Arizona, Mexico
     Z   W   P

Sciurus nelsoni   Nelson's squirrel
O   C. Mexico

Sciurus niger   Eastern fox squirrel
O   C.and E. United States, N.E.Mexico
     Z   W   P

Sciurus oculatus   Peter's squirrel
O   C. I Mexico

Sciurus poliopus   Mexican gray squirrel
O   S.Mexico

Sciurus richmondi   Richmond's squirrel
O   Nicaragua

Sciurus sinaloensis   Sinaloan squirrel
O   Mexico,(Sinaloa)

Sciurus socialis   Sociable squirrel
O   Mexico.(Guerrero to Oaxaca)

Sciurus truei   Sonoran squirrel
O   Mexico,(Sonora)

Sciurus variegatoides
O   Mexico to Panama

Variegated squirrel

Sciurus vulgaris
O   Europe to Japan
Z   W   P

Red squirrel

Sciurus yucatanensis
O   Yucatan, Guatemala

Yucatan squirrel

Spermophilopsis leptodactylus
O   S.Russia, Afghanistan Iran

Long clawed ground squirrel

Spermophilus adocetus
O   Mexico

Tropical ground squirrel

Spermophilus annulatus
O   Mexico

Ring tailed ground squirrel

Spermophilus armatus
O   Montana,Wyoming,Utah,Idaho
Z   W   P

Uinta ground squirrel

Spermophilus atricapillus
O   Baja California

Rock squirrel

Spermophilus beecheyi
O   Oregon, California
Z   W   P

California ground squirrel

Spermophilus beldingi
O   Oregon,N.E.California,Nevada
Z   W   P

Beldings ground squirrel

Spermophilus brunneus
O   Idaho
Z   W   P

Idaho ground squirrel

Spermophilus citellus
O   Germany to Manchuria

European souslik

Spermophilus columbianus
O   British Columbia to Idaho
Z   W   P

Columbian ground squirrel

Spermophilus franklinii
O   Canada, United States
Z   W   P

Franklins ground squirrel

Spermophilus fulvus
O   Afghanistan, Iran

Large toothed souslik

Spermophilus lateralis   Golden mantled ground squirrel
O   S.W.Canada, W.United States
     Z   W   P

Spermophilus madrensis   Sierra Madre mantled ground squirrel
O   Mexico,(Chihuahua)

Spermophilus major   Ground squirrel
O   Russia

Spermophilus mexicanus   Mexican ground squirrel
O   New Mexico, Texas to C. Mexico
     Z   W   P

Spermophilus mohavensis   Mohave ground squirrel
O   S.C. California
     Z   W   P

Spermophilus pallidicauda   Ground squirrel
O   Mongolia

Spermophilus parryii   Arctic ground squirrel
O   Siberia, Alaska, N.W.Canada
     Z   W   P

Spermophilus perotensis   Perote ground squirrel
O   E.C.Mexico

Spermophilus pygmaeus   Little souslik
O   S. Russia

Spermophilus richardsonii   Richardson's squirrel
O   Alberta, Nevada to Minnesota and Colorado
     Z   W   P

Spermophilus saturatus   Cascade golden mantled ground squirrel
O   British Columbia, Washington
     Z   W   P

Spermophilus spilosoma   Spotted ground squirrel
O   North Dakota, Utah to Mexico
     Z   W   P

Spermophilus suslicus   Spoted souslik
O   Poland to the Volga river

Spermophilus tereticaudus   Round tailed ground squirrel
O   Nevada, California, Arizona, N.W.Mexico
     Z   W   P

Spermophilus townsendii   Towsend's ground squirrel
O   W. United States
     Z   W   P

Spermophilus tridecemlineatus   Thirteen lined ground squirrel
O   Alberta to Ohio and S. Texas
    Z   W   P

Spermophilus undulatus   Siberian souslik
O   S. Siberia, Mongolia

Spermophilus variegatus   Rock squirrel
O   S.W. United States to C. Mexico
    Z   W   P

Spermophilus washingtoni   Washington ground squirrel
O   Washington, Oregon
    Z   W   P

Sundasciurus brookei   Brooke's squirrel
O   Borneo

Sundasciurus hippurus   Horse tailed squirrel
O   Malaya , Sumatra

Sundasciurus lowii   Low's squirrel
O   Malaya to Borneo

Sundasciurus mindanensis   Sunda tree squirrel
O   Philippines (Mindanao)

Sundasciurus mollendorffi   Calamian squirrel
O   Philippines,(Culion Isl)

Sundasciurus philippinensis   Mindanao squirrel
O   Philippines,(Mindanao)

Sundasciurus steeri   Palawan squirrel
O   Philippines,(Palawan)

Sundasciurus tenius   Slender squirrel
O   Thailand to Borneo

Syntheosciurus brochus   Mountain squirrel
O   Panama

Syntheosciurus poasensis   Poas mountain squirrel
O   Costa Rica

Tamias striatus   Eastern chipmunk
O   S.E.Canada and E. United States
    Z   W   P

Tamiasciurus douglasii   Douglas squirrel
O   British Columbia to Baja California
    Z   W   P

Tamiasciurus hudsonicus   Red squirrel
O   Alaska, Canada, United States
    Z   W   P

Tamiops mac clellandii   Himalayan striped squirrel
O   Himalaya to Indochina

Tamiops maritimus   Striped tree squirrel
O   China,Laos, Viet Nam

Tamiops rodolphei   Striped tree squirrel
O   Indochina

Tamiops swinhoei   Asiatic striped squirrel
O   China, Indochina, Taiwan

Xerus erythropus   Western ground squirrel
O   Senegal to Kenya

Xerus inaurius   Bristly ground squirrel
O   S. Africa

Xerus princeps   Namibia ground squirrel
O   Angola, Namibia

Xerus rutilus   African ground squirrel
O   Sudan to Tanzania

Aeretes melanopterus   Groove toothed ground flying squirrel
O   N. China

Aeromys terphromelas   Large black flying squirrel
O   Malaya

Belomys pearsoni   Hairy footed flying squirrel
O   S.Asia

Eupetaurus cinereus   Woolly flying squirrel
O   Pakistan

Glaucomys sabrinus   Northern flying squirrel
O   Alaska, Canada, W.United States

Glaucomys volans   Southern flying squirrel
O   S.E.Canada to Honduras

Hylopetes alboniger   Flying squirrel
O   Nepal to Indochina

Hylopetes fimbriatus   Flying squirrel
O   Afghanistan to India

Hylopetes harrisoni O Borneo	Flying squirrel
Hylopetes lepidus O S.E.Asia	Gray cheeked flying squirrel
Hylopetes nigripes O Philippines,(Palawan Isl,)	Palawan flying squirrel
Hylopetes phayrei O Burma, Thailand, Laos	Phayre's flying squirrel
Hylopetes sagitta O Java	Java flying squirrel
Hylopetes spadiceus O S.E.Asia	Red cheeked flying squirrel
Iomys horsfieldi O Malaya, Borneo	Horsfield's flying squirrel
Petaurillus emiliae O Borneo,(sarawak)	Pygmy flying squirrel
Petaurillus hosie O Borneo (Sarawak)	Pygmy flying squirrel
Petaurillus kinlochii O Malaya	Selangor pygmy flying squirrel
Petaurista alborufus O Burma to Taiwan	Giant flying squirrel
Petaurista elegans O Nepal to Borneo	Spotted giant flying squirrel
Petaurista leucogenys O Japan, China	Japan giant flying squirrel
Petaurista magnificus O Nepal, Sikkim	Hodgson's flying squirrel
Petaurista petaurista O S.and S.E.Asia	Red giant flying squirrel
Petinomys crinitus O Philippines,(Basilan Isl,)	Basilan flying squirrel
Petinomys electilis O Hainan	Pygmy flying squirrel
Petinomys fuscocapillus O India, Sri Lanka	Small flying squirrel

159

Petinomys genibarbis
O   Malaya to Borneo

Whiskered flying squirrel

Petinomys hageni
O   Sumatra, Borneo

Hagen's flying squirrel

Petinomys setosus
O   Malaya to Borneo

Temminck's pygmy flying squirrel

Petinomys vordermanni
O   Malaya

Vordermann's flying squirrel

Pteromys momonga
O   Japan

Japanese flying squirrel

Pteromys volans
O   Finland to Japan

European flying squirrel

Pteromyscus pulverulentus
O   S.E.Asia

Smoky flying squirrel

Trogopterus xanthipes
O   China

Complex toothed flying squirrel

FAMILY : GEOMYIDAE, Pocket gophers

Pocket gophers are the most highly fossorial to North American rodents. They have small eyes, small externals ears, and short tails. Moderately small, they weigh from 4 oz. to 2 lbs. External fur lined cheek pouches are present.

Cratogeomys fumosus
O   Mexico, (Colima)

Smoky pocket gopher

Cratogeomys gymnurus
O   Mexico (Jalisco)

Loano pocket gopher

Cratogeomys merriami
O   C. Mexico

Meriam's pocket gopher

Cratogeomys neglectus
O   Mexico

Queretaro pocket gopher

Cratogeomys oreocetes
O   Mexico

Pocket gopher

Cratogeomys pergrinus
O   Mexico

Pocket gopher

Cratogeomys perotensis
O   Mexico,(Veracruz)

Perote pocket gopher

Cratogeomys planiceps
O   Mexico

Pocket gopher

Cratogeomys tylorhinus   Taylor's pocket gopher
O   Mexico,(Hidalgo)

Cratogeomys zinseri   Zinser's pocket gopher
O   Mexico,(Jalisco)

Geomys arenarius   Desert pocket gopher
O   New, Mexico, Texas, Mexico.

Geomys bursarius   Plains pocket gopher
O   C. United States

Geomys colonus   Colonial pocket gopher
O   Georgia

Geomys fontanelus   Shermans pocket gopher
O   Georgia

Geomys personatus   Texas pocket gopher
O   Texas, Mexico

Geomys pinetis   Southern pocket gopher
O   Alabama, Georgia, Florida

Heterogeomys hispidus   Hispid pocket gopher
O   Mexico

Heterogeomys lanius   Big pocket gopher
O   Mexico, (Veracruz)

Macrogeomys cavator   Chiriqui pocket gopher
O   Panama (Chiriqui), Costa Rica

Macrogeomys cherriei   Cherries pocket gopher
O   Costa Rica

Macrogeomys dariensis   Darien pocket gopher
O   Panama

Macrogeomys heterodus   Variable pocket gopher
O   Costa Rica

Macrogeomys matagalpae   Nicaraguan pocket gopher
O   Nicaragua,Honduras

Macrogeomys underwoodi   Underwoods pocket gopher
O   Costa Rica

Orthogeomys cuniculus   Oaxacan pocket gopher
O   Mexico,(Oaxaca)

Orthogeomys grandis   Large pocket gopher
O   Guerrero to Honduras

Orthogeomys pygacanthus   El Salvador pocket gopher
O   El Salvador

Pappogeomys alcorni   Alcorn's pocket gopher
O   Mexico,(Jalisco)

Pappogeomys bulleri   Buller's pocket gopher
O   Mexico, (Jalisco)

Pappogeomys castanops   Yellow faced pocket gopher
O   Kansas, Colorado to C. Mexico

Thomomys baileyi   Bailey's pocket gopher
O   Arizona, Texas, New Mexico

Thomomys bottae   Bottas pocket gopher
O   S.W.United States to Mexico

Thomomys bulbivorous   Giant pocket gopher
O   Oregon

Thomomys idahoensis   Idaho pocket gopher
O   Idaho

Thomomys mazama   Mazama pocket gopher
O   Oregon

Thomomys monticola   Sierra pocket gopher
O   Oregon, California, Nevada

Thomomys talpoides   Northern pocket gopher
O   S.W.Canada, N.W. United States

Thomomys townsendii   Townsend's pocket gopher
O   California, Nevada, Idaho

Thomomys umbrinus   Southern pocket gopher
O   Arizona, New Mexico, Mexico

Zygogeomys trichopus   Tuza
O   Mexico,(Michoacan)

FAMILY: HETEROMYIDAE,   Kangaroo rats   & pocket mice

Dwellers in generally arid and semi- arid regions of the New World these rodents show a tendency toward saltation (jumping) as a primary mode of progression. Many have a remarkable ability to survive for long periods on dry seeds and no free water.

Dipodomys agilis   Pacific kangaroo rat
O   S.W.California, Baja California

Dipodomys deserti   Desert kangaroo rat
O   Nevada, California, Arizona, N.Mexico

Dipodomys elator
O   Oklahoma, Texas
Texas kangaroo rat (R)

Dipodomys elephantinus
O   California (San Benito)
Big eared kangaroo rat

Dipodomys gravipes
O   N.W.Baja California
San Quintin kangaroo rat

Dipodomys heermanni
O   California
Heermann kangaroo rat (E)

Dipodomys ingens
O   W.C. California
Giant kangaroo rat (E*)

Dipodomys insularis
O   Baja California,(San Jose Isl,)
San Jose kangaroo rat

Dipodomys margaritae
O   Baja California,(Margarita Isl,)
Margarita kangaroo rat

Dipodomys merriami
O   S.W.United States, N.Mexico
Merriam kangaroo rat

Dipodomys microps
O   Oregon, California,Arizona, Nevada
Great Basin kangaroo rat

Dipodomys nelsoni
O   N.C.Mexico
Nelson's kangaroo rat

Dipodomys nitratoides
O   C. California
Fresno kangaroo rat (E*)

Dipodomys ordii
O   Alberta, Saskatchewan, W.United States,Mexico
Ord's kangaroo rat

Dipodomys panamintinus
O   Nevada, California
Panamint kangaroo rat

Dipodomys paralius
O   Baja California,(Santa Catarina Isl,)
Kangaroo rat

Dipodomys peninsularis
O   Baja California
Baja California kangaroo rat

Dipodomys phillipsii
O   C. Mexico
Phillip's kangaroo rat

Dipodomys spectabilis
O   Arizona, New Mexico, Texas,N. Mexico
Banner tailed kangaroo rat

Dipodomys stephensi
O   S.W.California
Stephen's kangaroo rat

Dipodomys venustus   Santa Cruz kangaroo rat
O   Coastal California from San Francisco Bay to Morro Bay

Heteromys australis   Spiny pocket mouse
O   Panama to Ecuador

Heteromys desmarestianus   Forest spiny pocket mouse
O   Mexico to Colombia

Heteromys gaumeri   Gaumer's spiny pocket mouse
O   Yucatan, Guatemala

Heteromys goldmani   Goldman's spiny pocket mouse
O   S.Mexico, Guatemala

Heteromys lepturus   Spiny pocket mouse
O   Mexico,(Oaxaca)

Heteromys longicaudatus   Long tailed spiny pocket mouse
O   Mexico,(Tabasco)

Heteromys nelsoni   Nelson's spiny pocket mouse
O   Mexico,(Chiapas)

Heteromys nigricaudatus   Goodwin's spiny pocket mouse
O   Mexico,(Oaxaca)

Heteromys oresterus   Mountain spiny pocket mouse
O   Costa Rica

Heteromys temporalis   Motzoronga spiny pocket mouse
O   Mexico,(Veracruz)

Liomys adspersus   Panama spiny pocket mouse
O   Panama

Liomys annectens   Pluma spiny pocket mouse
O   Mexico,(Oaxaca)

Liomys anthonyi   Anthony's spiny pocket mouse
O   Guatemala

Liomys bulleri   Spiny pocket mouse
O   Mexico,(Jalisco)

Liomys crispus   Rough spiny pocket mouse
O   Mexico,(Chiapas)

Liomys guerrerensis   Spiny pocket mouse
O   Mexico,(Guerrero)

Liomys heterothrix   Honduran spiny pocket mouse
O   Honduras

Liomys irroratus O Mexico,Texas	Mexican spiny pocket mouse
Liomys pictus O W.and S.Mexico	Painted spiny pocket mouse
Liomys pinetorum O Mexico,(Chiapas)	Chiapan spiny pocket mouse
Liomys salvini O Guatemala to Costa Rica	Salvin's spiny pocket mouse
Liomys spectabilis O Mexico,(Jalisco)	Jalisco spiny pocket mouse
Microdipodops megacephalus O Great basin region of W. United States	Dark kangaroo mouse
Microdipodops pallidus O S.W.Nevada,California	Pale kangaroo mouse
Perognathus alticola O S.C.California	White eared pocket mouse
Perognathus amplus O Arizona, N.W.New Mexico	Arizona pocket mouse
Perognathus anthonyi O Baja California ,(Cerros Isl.)	Anthony's pocket mouse
Perognathus apache O Colorado,Utah , Arizona, New Mexico	Apache pocket mouse
Perognathus arenarius O Baja California	Little desert pocket mouse
Perognathus artus O N.W.Mexico	Narrow skulled pocket mouse
Perognathus californicus O C.California to Baja California	California pocket mouse
Perognathus fallax O S. California to Baja California	San Diego pocket mouse
Perognathus fasciatus O Alberta,Manitoba to Colorado	Olive backed pocket mouse
Perognathus flavescens O Minnesota to N.C.Mexico	Plains pocket mouse
Perognathus flavus O S.Dakota to N.Mexico	Silky pocket mouse

165

Perognathus formosus
O   Nevada,Utah to Baja California

Long tailed pocket mouse

Perognathus goldmani
O   Mexico

Goldman's pocket mouse

Perognathus hispidus
O   North Dakota to C.Mexico

Hispid pocket mouse

Perognathus inornatus
O   San Joaquin valley of California

San Joaquin pocket mouse

Perognathus intermedius
O   S.W.United States,N.Mexico

Rock pocket mouse

Perognathus lineatus
O   Mexico

Lined pocket mouse

Perognathus longimembris
O   Oregon and Utah to N.W.Mexico

Little pocket mouse

Perognathus merriami
O   Texas, N.Mexico

Merriam's pocket mouse

Perognathus nelsoni
O   New Mexico, Texas, N.C.Mexico

Nelson's pocket mouse

Perognathus parvus
O   British Columbia to California and Utah

Great basin pocket mouse

Perognathus penicillatus
O   S.W.United States to Mexico

Desert pocket mouse

Perognathus pernix
O   Mexico,(Sinaloa)

Sinaloan pocket mouse

Perognathus spinatus
O   Baja California

Spiny pocket mouse

Perognathus xanthonotus
O   California,(Kern county)

Yellow eared pocket mouse

FAMILY: CASTORIDAE,   Beavers

Beavers are large, semi-aquatic rodents, inhabiting inland waterways of the United States, Canada, Alaska, northern Europe, and northern Asia. Well adapted for aquatic life, the body is insulated by fine underfur, the large hind feet are webbed, the tail is broad, flat and largely hairless, the eyes have a fully developed nictitating membrane, the nostrils and ear opening can be closed. Beavers are remarkable builders and modify their environment with systems of canals and dams.

Castor canadensis   Beaver
O   Alaska, Canada, United States, N. Mexico
    Z   W   P.W
                                               P.Z

Castor fiber   European beaver
O   Palaearctic region of Europe and Asia
    Z   W   P.W
                                               P.Z

FAMILY: ANOMALURIDAE, Scaly tailed squirrel

Although these animals resemble squirrels in external appearance, their internal anatomy indicate no close relationship to squirrels. Most of them are nocturnal

Anomalurus beecrofti   Beecroft's scaly tailed squirrel
O   W.and C.Africa

Anomalurus cinereus   Tanzania flying squirrel
O   Tanzania

Anomalurus derbianus   Lord Derby's flying squirrel
O   Sierra Leone to Mozambique

Anomalurus jacksoni   Flying squirrel
O   E.Africa to Angola

Anomalurus orientalis   Flying squirrel
O   Zanzibar

Anomalurus peli   Pel's flying squirrel
O   Sierra Leone to Ghana

Anomalurus pusillus   Little flying squirrel
O   Gabon, Cameroon, Zaire

Idiurus langi   Pygmy scaly tailed squirrel
O   Zaire

Idiurus macrotis   Long eared flying squirrel
O   Sierre Leone to Zaire

Idiurus panga   Flying squirrel
O   Zaire

Idiurus zenkeri   Zinker's flying squirrel
O   Cameroon to Zaire

Zenkerella insignis   Non flying scaly tailed squirrel
O   W.Africa

FAMILY: PEDETIDAE, Springhare

The Springhare is terrestrial and fossorial; digs burrows for shelter and protection; inhabits sandy soils in arid country.

Pedetes capensis   Springhare
O   Zaire to Kenya to South Africa

MYOMORPHS

FAMILY:CRICETIDAE,Deermice,Woodrats,Lemmings,Voles,Muskrats, Gerbils, and Hamsters

This is an extremely varied family. Generally "mouse-shaped" , They range in size from 1/4 oz. to 3 lbs. It is the largest family in number of species.

Akodon affinis   Field mouse
O   Colombia

Akodon albiventer   Field mouse
O   Peru, Bolivia,Chile,Argentina

Akodon amoenus   Field mouse
O   Peru

Akodon andinus   Field mouse
O   The Andes from Peru to Chile and Argentina

Akodon arviculoides   Field mouse
O   Paraguay, Brazil

Akodon azarae   Field mouse
O   Argentina,Uruguay

Akodon berlepschii   Field mouse
O   Bolivia, Peru,Chile

Akodon bogotensis   Field mouse
O   The Andes of Peru, Bolivia, Argentina

Akodon boliviensis   Field mouse
O   S.W.Peru to Argentina

Akodon budini   Field mouse
O   Mountains of N.W.Argentina

Akodon caenosus   Field mouse
O   Argentina,Bolivia

Akodon chacoensis   Field mouse
O   N.Argentina

168

Akodon dolores   Field mouse
O   C. Argentina

Akodon iheringi   Field mouse
O   S.tip of Brazil

Akodon illuteus   Field mouse
O   N.W..Argentina

Akodon iniscatus   Field mouse
O   C. Argentina

Akodon jelskii   Field mouse
O   The Andes from Peru to Argentina

Akodon kempi   Field mouse
O   Islands of Parana river

Akodon lactens   Field mouse
O   Mountains of N.W.Argentina

Akodon lanosus   Field mouse
O   S. Argentina and Chile

Akodon lasiotis   Field mouse
O   E.Brazil

Akodon latebricola   Field mouse
O   The Andes of Ecuador

Akodon longipilis   Field mouse
O   Chile, Argentina,(Tierra del Fuego)

Akodon mimus   Field mouse
O   The Andes of S.W.Peru

Akodon mollis   Field mouse
O   The Andes from Ecuador to Bolivia

Akodon nigritta   Field mouse
O   E.and S. Brazil

Akodon obscurus   Field mouse
O   Argentina, Uruguay

Akodon olivaceus   Field mouse
O   Argentina,Chile

Akodon orophilus   Field mouse
O   Peru

Akodon pacificus   Field mouse
O   Mountains of W. Bolivia

Akodon puer   Field mouse
O   Peru, Bolivia

Akodon sanborni   Field mouse
O   Chile

Akodon serrensis   Field mouse
O   E.Brazil

Akodon surdus   Field mouse
O   Mountains of S.E.Peru

Akodon tapirapoanus   Field mouse
O   Bolivia, Brazil

Akodon urichi   Field mouse
O   The Andes from Venezuela to Bolivia,Trinidad

Akodon varius   Field mouse
O   Argentina, Bolivia,Paraguay

Akodon xanthorhinus   Field mouse
O   Chile, Argentina

Andinomys edax   Andean mouse
O   Bolivia, Argentina,Peru

Anotomys leander   Ecuador fish eating rat
O   N.Ecuador

Baiomys musculus   Small pygmy mouse
O   Mexico to Nicaragua

Baiomys taylori   Northern pygmy mouse
O   Arizona,Texas to Mexico

Blarinomys breviceps   Brazilian shrew mouse
O   E.Brazil

Calomys callosus   Vesper mouse
O   Bolivia, Argentina,Paraguay

Calomys dubius   Vesper mouse
O   Uruguay, Argentina

Calomys expulsus   Vesper mouse
O   Brazil

Calomys frida   Vesper mouse
O   Peru

Calomys gracilipes   Vesper mouse
O   Argentina

Calomys laucha   Vesper mouse
O   Argentina

Calomys pepidus   Vesper mouse
O   The Andes of Peru, Bolivia, Chile,Argentina

Calomys muriculus   Vesper mouse
O   Bolivia

Calomys tener   Vesper mouse
O   Brazil

Calomys venustus   Vesper mouse
O   Argentina

Calomyscus bailwardi   Mouse like hamster
O   Iran, Afghanistan,Pakistan

Chilomys instans   Colombian forest mouse
O   Colombia,Ecuador,Venezuela

Chinchillula sahamae   Altiplano chinchilla mouse
O   Highlands of Peru, Bolivia, Argentina, Chile

Cricetulus alticola   Short tailed tibetan hamster
O   Tibet, Pakistan

Cricetulus barabensis   Striped hamster
O   N.China to Siberia

Cricetulus eversmanni   Evermann's hamster
O   Russia,(Kazakh)

Cricetulus lama   Tibetan hamster
O   Tibet

Cricetulus longicaudatus   Long tailed hamster
O   Mongolia,Siberia,China

Cricetulus migratorius   Migratory hamster
O   S.E. Europe to Iran and Mongolia

Cricetulus triton   Long tailed hamster
O   China, Korea, Siberia,

Cricetus cricetus   **Common hamster**
O   Europe,W. Asia

Daptomys venezuelae   Aquatic rat
O   N. Venezuela

Eligmodontia typus   Desert mouse
O   Argentina, Chile,Peru

171

Euneomys chinchilloides
O   Tierra Del Fuego

Patagonian chinchilla mouse

Euneomys fossor
O   N.W. Argentina

Chinchilla mouse

Euneomys mordax
O   W.Argentina

Chinchilla mouse

Euneomys petersoni
O   S.Argentina

Chinchilla mouse

Galenomys garleppi
O   Bolivia,Peru

Mouse

Holochilus brasiliensis
O   Venezuela to Argentina

Marsh rat

Holochilus magnus
O   Uruguay, Brazil,Argentina

Marsh rat

Ichthyomys hydorbates
O   Venezuela to Ecuador

Fish eating rat

Ichthyomys stolzmanni
O   Ecuador, Peru

Fish eating rat

Irenomys tarsalis
O   Chile, Argentina

Chilean rat

Lenoxus apicalis
O   Peru, Brazil

Peruvian rat

Megalomys audreyae
O   Lesser Antilles

Audrey's muskrat

Megalomys luciae
O   Lesser Antilles

Santa Lucia muskrat

Mesocricetus auratus
O   Europe,Asia

Golden hamster

Myospalax fontanierii
O   China

Chinese zokor

Myospalax myospalax
O   Siberia, Manchuria

Altai zokor

Myospalax psilurus
O   Siberia,N.E. China

Manchurian zokor

Mystromys albicaudatus
O   S.Africa

White tailed rat

172

Mystromys longicaudatus
O   Tanzania

White tailed rat

Neacomys guianae
O   Guyana, Venezuela

Bristly mouse

Neacomys pictus
O   Panama to Ecuador

Painted bristly mouse

Neacomys pusillus
O   Colombia, Ecuador

Bristly mouse

Neacomys spinosus
O   Colombia to Peru

Bristly mouse

Nectomys alfari
O   Panama to Ecuador

Alfaros water rat

Nectomys squamipes
O   N. and C.South America

Water rat

Nelsonia neotomodon
O   C. Mexico

Diminuitive wood rat

Neotoma albigula
O   S.W.United States to C.Mexico

White throated woodrat

Neotoma alleni
O   S.W.coast Mexico

Allen's woodrat

Neotoma angustapalata
O   Mexico,(Tamaulipas)

Woodrat

Neotoma anthonyi
O   Baja California(Todos Santas Isl.)

Woodrat

Neotoma bryanti
O   Baja california (Cedros Isl)

Woodrat

Neotoma bunkeri
O   Baja California (Coronados Isl.)

Woodrat

Neotoma chrysomelas
O   Nicaragua to Honduras

Nicaraguan woodrat

Neotoma cinerea
O   British Columbia to Arizona

Bushy tailed woodrat

Neotoma floridana
O   South Dakota, New York to Texas and Florida

Eastern woodrat

Neotoma fuscipes
O   Oregon to Baja California

Dusky footed woodrat

173

Neotoma goldmani   Goldman's woodrat
O   N.C.Mexico

Neotoma lepida   Desert woodrat
O   Oregon to Baja California

Neotoma martinsensis   Woodrat
O   Baja California (San Martin Isl.)

Neotoma mexicana   Mexican woodrat
O   Colorado to Honduras

Neotoma micropus   Southern plains woodrat
O   S.Kansas to New Mexico

Neotoma nelsoni   Nelson's woodrat
O   Mexico,(Veracruz)

Neotoma palatina   Bolanos woodrat
O   Mexico,(Jalisco)

Neotoma phenax   Sonoran woodrat
O   Mexico,(Sonora)

Neotoma stephensi   Stephen's woodrat
O   Utah, Arizona, New Mexico

Neotoma varia   Woodrat
O   Mexico,(Turner's Isl)

Neotomodon alstoni   Volcano mouse
O   Mexico

Neotomys ebriosus   Andean swamp rat
O   Peru to Argentina

Neusticomys monticolus   Fish eating mouse
O   N.Ecuador

Notiomys angustus   Long clawed mouse
O   W.Argentina

Notiomys delfini   Long clawed mouse
O   Chile,(Magallan)

Notiomys edwardsii   Long clawed mouse
O   S.tip Argentina

Notiomys macronyx   Long clawed mouse
O   Argentina, Chile

Notiomys megalonyx   Long clawed mouse
O   Chile

Notiomys valdivianus   Long clawed mouse
O   Chile,Argentina

Ochrotomys nuttalli   Golden mouse
O   S.E.United States

Onychomys leucogaster   Northern grasshopper mouse
O   Saskatchewan to Texas

Onychomys torridus   Southern grasshopper mouse
O   S.W.United States to C.Mexico

Oryzomys albigularis   Rice rat
O   Venezuela to Bolivia

Oryzomys alfaroi   Alfaro's rice rat
O   Mexico to Ecuador

Oryzomys altissimus   Rice rat
O   Ecuador, Peru

Oryzomys andinus   Rice rat
O   N.Peru

Oryzomys antillarum   Jamaican rice rat
O   Jamaica

Oryzomys aphrastus   Harris rice rat
O   Costa Rica,(San Jose)

Oryzomys arenalis   Rice rat
O   N.E.Peru

Oryzomys azueremsis   Azuero rice rat
O   Panama,(Veraguas)

Oryzomys balneator   Rice rat
O   Ecuador

Oryzomys bicolor   Rice rat
O   Guyana to Bolivia

Oryzomus bombycinus   Silky rice rat
O   Costa rica to Ecuador

Oryzomys borreroi   Rice rat
O   N. Colombia

Oryzomys buccinatus   Rice rat
O   Paraguay, Argentina

Oryzomys caliginosus   Dusky rice rat
O   Panama to Ecuador

175

Oryzomys capito   Rice rat
O   Venezuela to Paraguay

Oryzomys catherinae   Rice rat
O   Guyana,Brazil

Oryzomys caudatus   Rice rat
O   Mexico,(Oaxaca)

Oryzomys chaparensis   Rice rat
O   Bolivia

Oryzomys couesi   Coue's rice rat
O   Mexico to Costa Rica

Oryzomys cozumelae   Cozumel rice rat
O   Mexico,(Cozumel Isl.)

Oryzomys darwini   Santa Cruz rice rat
O   Mexico,(Santa Cruz Isl.)

Oryzomys delicatus   Rice rat
O   Colombia to Guyana

Oryzomys delticola   Rice rat
O   Argentina,Uruguay

Oryzomys devius   Boquete rice rat
O   Panama,(Chiriqui)

Oryzomys endersi   Ender's rice rat
O   Panama,(Barro Colorado Isl,)

Oryzomys flavescens   Rice rat
O   Argentina, Uruguay

Oryzomys fulgens   Thomas rice rat
O   Mexico

Oryzomys fulvescens   Pygmy rice rat
O   Mexico to Venezuela

Oryzomys galapagoensis   Rice rat
O   Galapagos,(San Cristobal Isl.)

Oryzomys gatunensis   Gatun rice rat
O   Panama,(canal zone)

Oryzomys hammondi   Rice rat
O   Ecuador

Oryzomys indefessus   Rice rat
O   Galapagos,(Santa Cruz Isl.)

Oryzomys intectus   Rice rat
O   C.Colombia

Oryzomys longicaudatus   Rice rat
O   Peru to Patagonia

Oryzomys macconnelli   Rice rat
O   Guyana to Colombia

Oryzomys maculipes   Rice rat
O   E.Brazil

Oryzomys mamorae   Rice rat
O   Ecuador to Bolivia

Oryzomys marmosurus   Rice rat
O   Guyana to Colombia

Oryzomys melanostoma   Rice rat
O   Peru

Oryzomys melanotis   Black eared rice rat
O   Mexico to El Salvador

Oryzomys melleus   Rice rat
O   S.E.Ecuador

Oryzomys microtis   Rice rat
O   Amazonian Brazil

Oryzomys mincae   Rice rat
O   N.Colombia

Oryzomys minutus   Rice rat
O   Colombia to Peru

Oryzomys munchiquensis   Rice rat
O   W.Colombia

Oryzomys narboroughi   Rice rat
O   Galapagos,(Fernandina Isl.)

Oryzomys nelsoni   Rice rat
O   Mexico,(Maria Madre Isl.)

Oryzomys nigripes   Rice rat
O   Paraguay, Brazil

Oryzomys osgoodi   Rice rat
O   Peru,Ecuador

Oryzomys palustris   Rice rat
O   S.E. United States,Mexico to Panama

Oryzomys peninsulae   Rice rat
O   Baja California

Oryzomys phaeotis   Rice rat
O   S.E.Peru

Oryzomys polius   Rice rat
O   N.Peru

Oryzomys ratticeps   Rice rat
O   Paraguay,Brazil

Oryzomys roberti   Rice rat
O   S.E.Brazil

Oryzomys robustulus   Rice rat
O   Ecuador

Oryzomys simplex   Rice rat
O   E.Brazil

Oryzomys spodiurus   Rice rat
O   Ecuador

Oryzomys subflavus   Rice rat
O   E.Brazil

Oryzomys talamaccae   Talamancan rice rat
O   Costa Rica to Panama

Oryzomys tectus   Panama rice rat
O   Panama

Oryzomys trabeatus   Regal rice rat
O   Panama,(Rio Jasucito)

Oryzomys trinitatis   Rice rat
O   Venezuela, Colombia

Oryzomys utiaritensis   Rice rat
O   N.Brazil

Oryzomys victus   Rice rat
O   Lesser Antilles,(St. Vincent)

Oryzomys villosus   Rice rat
O   N.Colombia

Oryzomys xantheolus   Rice rat
O   Peru

Oryzomys zunigae   Rice rat
O   Peru

Otonyctomys hatti   Yucatan vesper rat
O   Mexico,(Yucatan)

Ototylomys phyllotis   Big eared climbing rat
O   Yucatan to Costa Rica

Oxymycterus akodontius   Burrowing mouse
O   N.W. Argentina

Oxymycterus angularis   Burrowing mouse
O   E.Brazil

Oxymycterus delator   Burrowing mouse
O   Paraguay

Oxymycterus hispidus   Burrowing mouse
O   Brazil,Argentina

Oxymycterus inca   Burrowing mouse
O   Bolivia to Peru

Oxymycterus paramensis   Burrowing mouse
O   S.Bolivia to S.E.Peru

Oxymycterus roberti   Burrowing mouse
O   E.Brazil

Oxymycterus rufus   Burrowing mouse
O   E.Brazil to Argentina

Peromyscus altilaneus   Deer mouse
O   Guatemala,(Todos Santos)

Peromyscus attwateri   Texas mouse
O   Kansas to Arkansas and Texas

Peromyscus aztecus   Deer mouse
O   Mexico to Honduras

Peromyscus banderanus   Michoacan deer mouse
O   W. Coast Mexico

Peromyscus boyli   Brush mouse
O   California and Texas to Guatemala

Peromyscus bullatus   Perote mouse
O   Mexico,(Veracruz)

Peromyscus californicus   California mouse
O   C. California to Baja California

Peromyscus caniceps   Burt's deer mouse
O   Baja California,(Monserrate Isl.)

179

Peromyscus collatus   Canyon mouse
O   Mexico,(Turners Isl.)

Peromyscus comanche   Palo Duro mouse
O   Texas

Peromyscus crinitus   Canyon mouse
O   W. United States, Baja California

Peromyscus dickeyi   Dickey's deer mouse
O   Baja California,(Tortuga Isl.)

Peromyscus difficilis   Rock mouse
O   Colorado to S.C. Mexico

Peromyscus eremicus   Cactus mouse
O   S.W.United States,Mexico

Peromyscus evides   Deer mouse
O   W.Mexico

Peromyscus flavidus   Yellow deer mouse
O   Panama,(Chiriqui)

Peromyscus floridanus   Florida mouse
O   Florida

Peromyscus furvus   Blackish deer mouse
O   Mexico,(Veracurz)

Peromyscus gossypinus   Cotton mouse
O   S.E. United States

Peromyscus grandis   Big deer mouse
O   Guatemala,(Verapaz)

Peromyscus guardia   Angel Isl. deer mouse
O   San Lorenz and Mejia Isl, (gulf of California)

Peromyscus guatemalensis   Deer mouse
O   Guatemala,Honduras

Peromyscus gymnotis   Huehuetan deer mouse
O   Mexico,(Chiapas)

Peromyscus hondurensis   Deer mouse
O   Honduras

Peromyscus hylocetes   Wood mouse
O   C.Mexico

Peromyscus interparietalis   Deer mouse
O   Gulf of California (San Lorenz Isl.)

Peromyscus ixtlani   Deer mouse
O   Mexico

Peromyscus latirostris   Broadnosed deer mouse
O   Mexico,(San Louis Potosi)

Peromyscus lepturus   Slender tailed deer mouse
O   Mexico,(Oaxaca)

Peromyscus leucopus   White footed mouse
O   S. Canada to Yucatan

Peromyscus lophurus   Crested tailed mouse
O   Guatemala, El Salvador

Peromyscus maniculatus   Deer mouse
O   Alaska to Mexico

Peromyscus megalops   Brown deer mouse
O   Mexico,(Oaxaca)

Peromyscus mekisturus   Puebla deer mouse
O   Mexico,(Puebla)

Peromyscus melanocarpus   Deer mouse
O   Mexico,(Mt. Zempoaltepec)

Peromyscus melanophrys   Plateau mouse
O   Mexico

Peromyscus melanotis   Black eared mouse
O   C.Mexico

Peromyscus merriami   Merriam's mouse
O   N.W. Arizona ,Mexico

Peromyscus mexicanus   Mexican deer mouse
O   Veracruz to Costa Rica

Peromyscus nudipes   Naked footed deer mouse
O   Costa Rica

Peromyscus oaxacensis   Oaxacan deer mouse
O   Mexico,(Oaxaca)

Peromyscus ochraventer   El Carrizo deer mouse
O   Mexico,(Tamaulipas)

Peromyscus pectoralis   White ankled mouse
O   Oklahoma to Mexico

Peromyscus pembertoni   Pemberton's deer mouse
O   Gulf of California,(San Pedro Nolasca Isl.)

Peromyscus perfulvus   Marsh mouse
O   Mexico,(Jalisco)

Peromyscus pirrensis   Mt. Pirri deer mouse
O   Panama,(Mt. Pirri)

Peromyscus polionotus   Oldfield mouse
O   S.E. United States

Peromyscus polius   Chihuahuan mouse
O   Mexico,(Chihuahua)

Peromyscus pseudocrinitus   False canyon mouse
O   Baja California(Coronados Isl,)

Peromyscus sejugis   Island mouse
O   Gulf of California,(Santa Cruz, San Diego Isl.)

Peromyscus simulatus   Jico deer mouse
O   Mexico,(Veracruz)

Peromyscus sitkensis   Sitka mouse
O   Islands off Southeastern Alaska

Peromyscus slevini   Slevin's mouse
O   Baja California,(Santa Catalina Isl.)

Peromyscus stephani   Deer mouse
O   Baja California,(San Esteban Isl.)

Peromyscus stirtoni   Stirton's deer mouse
O   El Salvador,Guatemala,Honduras

Peromyscus thomasi   Thomas deer mouse
O   Mexico,(Guerrero)

Peromyscus truei   Pinon mouse
O   Oregon and Colorado to S.Mexico

Peromyscus yucatanicus   Yucatan deer mouse
O   Yucatan

Peromyscus zarhynchus   Chiapan deer mouse
O   Mexico,(Chiapas)

Phaenomys ferrugineus   Rio rice rat
O   E.Brazil

Phodopus roborovskii   Desert hamster
O   Mongolia,China

Phodopus sungorus   Hairy footed hamster
O   S.Siberia to Manchuria

182

Phyllotis amicus   Leaf eared mouse
O   Peru

Phyllotis andium   Leaf eared mouse
O   Ecuador to Peru

Phyllotis boliviensis   Leaf eared mouse
O   Bolivia,Peru

Phyllotis darwini   Leaf eared mouse
O   Chile, Peru,Argentina

Phyllotis gerbillus   Leaf eared mouse
O   N.W.Peru

Phyllotis griseoflavus   Leaf eared mouse
O   Argentina,Paraguay

Phyllotis haggardi   Leaf eared mouse
O   Ecuador

Phyllotis micropus   Leaf eared mouse
O   S.Argentina,S.Chile

Phyllotis osilae   Leaf eared mouse
O   N.W.Argentina to S.Peru

Phyllotis pictus   Leaf eared mouse
O   Peru, Bolivia

Phyllotis sublimis   Leaf eared mouse
O   S.Peru

Podoxymys roraimae   Mount Roraima mouse
O   Venezuela,Guyana, Brazil

Pseudoryzomys wavrini   Red nosed mouse
O   Paraguay, N.Argentina

Punomys lemminus   Puna mouse
O   S.E.Peru

Reithrodon physodes   Pampas gerbil
O   Uruguay to S.Argentina

Reithrodontomys brevirostris   Short nosed harvest mouse
O   Costa Rica, Nicaragua

Reithrodontomys burti   Short nosed harvest mouse
O   Mexico, (Sonora)

Reithrodontomys chrysopsis   Volcano harvest mouse
O   Mexico, (Michoacan)

Reithrodontomys creper   Chiriqui harvest mouse
O   W.Panama,Costa Rica

Reithrodontomys darienensis   Darien harvest mouse
O   Panama,(Darien)

Reithrodontomys fulvescens   Fulvous harvest mouse
O   S.C.United States to Nicaragua

Reithrodontomys gracilis   Slender harvest mouse
O   Yucatan to El Salvador

Reithrodontomys hirsutus   Hairy harvest mouse
O   Mexico,(Jalisco)

Reithrodontomys humulis   Eastern harvest mouse
O   S.E.United States

Reithrodontomys megalotis   Western harvest mouse
O   British Columbia to S.Mexico

Reithrodontomys mexicanus   Harvest mouse
O   S.Mexico to Ecuador

Reithrodontomys microdon   Small toothed harvest mouse
O   Mexico, Guatemala

Reithrodontomys montanus   Plains harvest mouse
O   Great plains region of the United States,Mexico

Reithrodontomys raviventris   Salt marsh harvest mouse (E)
O   San Francisco Bay area of California

Reithrodontomys rodriguezi   Harvest mouse
O   Costa Rica,(Cartago)

Reithrodontomys sumichrasti   Harvest mouse
O   S.Mexico to Panama

Reithrodontomys teniurostris   Narrow nosed harvest mouse
O   Guatemala

Rhagomys rufescens   Brazilian tree mouse
O   E.Brazil

Rheomys hartmanni   Water mouse
O   Panama,(Chiriqui)

Rheomys mexicanus   Water mouse
O   Mexico,(Oaxaca)

Rheomys raptor   Goldman's water mouse
O   Panama,(Mt. Pirri)

Rheomys thomasi   Thomas water mouse
O   Mexico to El Salvador

Rheomys trichotis   Water mouse
O   N. Colombia

Rheomys underwoodi   Water mouse
O   Costa Rica,Panama

Rhipidomys latimanus   Climbing mouse
O   Venezuela to Ecuador

Rhipidomys leucodactylus   Climbing mouse
O   Ecuador to N. Bolivia

Rhipidomys macconnelli   Climbing mouse
O   S.E.Venezuela

Rhipidomys mastacalis   Climbing mouse
O   Venezuela,Brazil

Rhipidomys scandens   Climbing mouse
O   Panama,(Mt.Pirri)

Rhipidomys sclateri   Climbing mouse
O   Guianas, Venezuela

Scapteromys aquaticus   Water rat
O   Argentina,(Isl. of Parana delta)

Scapteromys chacoensis   Water rat
O   N.Argentina

Scapteromys gnambiquarae   Water rat
O   Brazil

Scapteromys tomentosus   Water rat
O   Uruguay, Argentina

Scapteromys tumidus   Water rat
O   Uruguay, E.Argentina

Scolomys melanops   South American spiny mouse
O   South America

Scotinomys harrisi   Savannah brown mouse
O   Costa Rica,(Cartago)

Scotinomys longipilosus   Long furred brown mouse
O   Costa Rica,(Cartago)

Scotinomys teguina   Alston's brown mouse
O   Mexico to Panama

Scotinomys xerampelinus   Chiriqui brown mouse
O   Costa Rica to Panama

Sigmodon alleni   Allen's cotton rat
O   Mexico,(Jalisco)

Sigmodon alticola   Mountain cotton rat
O   Mexico,(Oaxaca)

Sigmodon arizonae   Arizona cotton rat
O   Arizona, W.Mexico

Sigmodon fulviventer   Tawny bellied cotton rat
O   Arizona,New Mexico,Mexico

Sigmodon guerrerensis   Guerreran cotton rat
O   Mexico,(Guerrero)

Sigmodon hispidus   Hispid cotton rat
O   Southern United States to Peru

Sigmodon leucotis   White eared cotton rat
O   Mexico,(Zacatecas)

Sigmodon macdougalli   Cotton rat
O   Mexico,(Oaxaca)

Sigmodon macrodon   Tehuantepec cotton rat
O   Mexico,(Oaxaca)

Sigmodon melanotis   Black eared cotton rat
O   Mexico,(Michoacan)

Sigmodon minimus   Least cotton rat
O   Arizona,New Mexico, Mexico

Sigmodon ochrognathus   Yellow nosed cotton rat
O   Arizona,New Mexico, Mexico

Sigmodon planifrons   Oaxacan cotton rat
O   Mexico,(Oaxaca)

Sigmodon vulcani   Volcano cotton rat
O   Mexico,(Fuego Jalisco)

Thomasomys aureus   Paramo mouse
O   Colombia,Peru

Thomasomys boeops   Paramo mouse
O   W.Ecuador

Thomasomys bombycinus   Paramo mouse
O   Colombia

186

Thomasomys cinereiventer   Paramo mouse
O   Colombia,Ecuador

Thomasomys cinereus   Paramo mouse
O   Ecuador, Peru

Thomasomys daphne   Paramo mouse
O   Bolivia, Peru

Thomasomys dorsalis   Paramo mouse
O   E.Brazil

Thomasomys gracilis   Paramo mouse
O   Ecuador, Peru

Thomasomys hylophilus   Paramo mouse
O   N.W.Colombia

Thomasomys incanus   Paramo mouse
O   Peru

Thomasomys ischyurus   Paramo mouse
O   Peru, Ecuador

Thomasomys kalinowskii   Paramo mouse
O   C.Peru

Thomasomys ladewi   Paremo mouse
O   N.W.Bolivia

Thomasomys laniger   Paramo mouse
O   Colombia, Venezuela

Thomasomys lugens   Paramo mouse
O   Venezuela to Ecuador

Thomasomys notatus   Paramo mouse
O   S.E.Peru

Thomasomys oenax   Paramo mouse
O   S.Brazil

Thomasomys oreas   Paramo mouse
O   Bolivia (Andes)

Thomasomys paramorum   Paramo mouse
O   Ecuador (Andes)

Thomasomys pictipes   Paramo mouse
O   N.E.Argentina

Thomasomys pyrrhonotus   Paramo mouse
O   Ecuador, Peru

187

Thomasomys rhoadsi   Paramo mouse
O   Ecuador

Thomasomys rosalinda   Paramo mouse
O   N.W.Peru

Thomasomys taczanowskii   Paramo mouse
O   N.W.Peru

Thomasomys vestitus   Paramo mouse
O   W.Venezuela

Tylomys bularis   Chiapan climbing rat
O   Mexico,(Tuxtla, Chiapas)

Tylomys fulviventer   Fulvous bellied climbing rat
O   Panama

Tylomys mirae   Climbing rat
O   Colombia, N.Ecuador

Tylomys nudicaudus   Peter's climbing rat
O   Chiapas to Nicaragua

Tylomys panamensis   Panama climbing rat
O   E.Panama

Tylomys tumbalensis   Tumbala climbing rat
O   Mexico,(Chiapas)

Tylomys watsoni   Watson's climbing rat
O   Panama

Wiedomys pyrrhorhinos   Wied's red nosed mouse
O   E.Brazil

Xenomys nelsoni   Magdalena rat
O   Mexico,(Colima)

Zygodontomys brevicauda   Cane rat
O   Colombia (Andes)

Zygodontomys cherriei   Cherrie's cane rat
O   Panama

Zygodontomys lasiurus   Cane rat
O   N.E.Brazil

Zygodontomys microtinus   Cane rat
O   Venezuela, Colombia

Zygodontomys punctulatus   Cane rat
O   Ecuador to Colombia

Brachytarsomys albicauda O Madagascar	Malagasy rat
Brachyuromys betsileoensis O Madagascar	Malagasy rat
Brachyuromys ramirohitra O Madagascar	Malagasy rat
Eliurus majori O Madagascar	Malagasy rat
Eliurus minor O Madagascar	Malagasy rat
Eliurus myoxinus O Madagascar	Malagasy rat
Eliurus tanala O Madagascar	Malagasy rat
Gymnuromys roberti O Madagascar	Malagasy rat
Hypogeomys animena O Madagascar	Malagasy rat
Macrotarsomys bastardi O Madagascar	Malagasy rat
Nesomys audeberti O Madagascar	Malagasy rat
Nesomys lambertoni O Madagascar	Malagasy rat
Nesomy rufus O Madagascar (Vohima)	Malagasy rat
Lophiomys imhause O E.Africa	Maned rat
Alticola macrotis O Mongolia, Siberia	High Mt. vole
Alticola roylei O Mongolia to India	High Mt. vole
Alticola stoliczkanus O Tibet, Kashmir, China	Flat skulled vole
Alticola strelzowi O Altai Mts,(Siberia)	Flat skulled vole

Arvicola richardsoni   Water vole
O   British Columbia to Utah

Arvicola terrestris   Water vole
O   Europe, Asia

Aschizomys lemminus   Vole
O   N.E.Siberia

Blanfordimys afghanus   Afghan vole
O   Afghanistan,S.Russia

Clethrionomys gapperi   Gapper's red backed mouse
O   Canada,N.United States

Clethrionomys glareolus   Bank vole
O   Europe,Asia

Clethrionomys occidentalis   California red backed mouse
O   Oregon, California

Clethrionomys rufocanus   Large toothed red backed mouse
O   Scandinavia to Siberia and Korea

Clethrionomys rutilus   Tundra red back vole
O   Scandinavia to the Bering Strait, Alaska

Dicrostonyx hudsonius   Hudson bay collared lemming
O   Quebec, Newfoundland

Dicrostonyx torquatus   Arctic lemming
O   Tundra region of Europe,Siberia,Alaska, Canada

Dolomys bogdanovi   Martinos snow vole
O   Yugoslavia

Ellobius fuscocapillus   Afghan mole vole
O   Iran,Afghanistan

Ellobius lutescens   Mole vole
O   Iran toTranscaucasia

Ellobius talpinus   Mole vole
O   S.Russia to Mongolia

Eothenomys chinensis   Pratt's vole
O   China

Eothenomys custos   Vole
O   China

Eothenomys fidelis   Vole
O   China

Eothenomys melanogaster   Pere David's vole
O   China, Taiwan,N.Burma

Eothenomys olitor   Vole
O   China

Eothenomys proditor   Vole
O   China

Hyperacrius fertilis   Kashmire vole
O   Kashmir

Hyperacrius wynnei   Murree vole
O   Kashmir

Lagurus curtatus   Sagebrush vole
O   N.W.United States,S.Canada

Lagurus lagurus   Steppe lemming
O   S.Russia, W.Siberia

Lagurus luteus   Yellow steppe lemming
O   China, Mongolia

Lemmus lemmus   Norway lemming
O   Norway, Sweden, Finland, Soviet Union

Lemmus nigripes   Black footed lemming
O   Alaska,(Pribilof Isl.)

Lemmus sibiricus   Brown lemming
O   Soviet Union,Alsaka to British Columbia

Microtus agrestis   Field vole
O   Europe to Siberia

Microtus arvalis   Common vole
O   Europe to Siberia and Asia Minor

Microtus bedfordi   Bedford's vole
O   China,(Kansu)

Microtus brandti   Brandt's vole
O   Siberia,Mongolia

Microtus cabrerae   Cabrera's vole
O   Spain

Microtus californicus   California vole
O   Oregon to Baja California

Microtus carruthesi   Carruther's vole
O   S.E.Russia,Turkestan

191

Microtus chrotorrhinus   Rock vole
O   Canada, Appalachians

Microtus clarkei   Clarke's vole
O   China, Burma

Microtus coronarius   Coronation Isl.vole
O   Alaska,(Coronation Isl.)

Microtus duodecimcostatus   Mediterranean pine vole
O   Spain to Greece

Microtus fortis   Reed vole
O   Siberia, China,Manchuria,Korea

Microtus gregalis   Singing vole
O   Russia, Asia

Microtus guatemelensis   Vole
O   Guatemala,(Huehuetenango)

Microtus gud   Vole
O   N.Asia Minor

Microtus hyperboreus   Vole
O   N.Siberia

Microtus irene   Vole
O   S. China to N.Burma

Microtus juldaski   Vole
O   Russia,Turkestan

Microtus kikuchii   Vole
O   Taiwan

Microtus leucurus   Blyth's vole
O   Tibet,China,Kashmir

Microtus longicaudus   Long tailed vole
O   Alaska to California and New Mexico

Microtus ludovicianus   Louisiana vole
O   E.Texas,S.W. Louisiana

Microtus mandarinus   Mandarin vole
O   N.China, Korea

Microtus mexicanus   Mexican vole
O   S.W.United States to C. Mexico

Microtus middendorffi   Middendorff's vole
O   Siberia

Microtus millicens   Szechuan vole
O   China,(Szechuan)

Microtus montanus   Mountain vole
O   British Columbia , Western United States

Microtus montebelli   Vole
O   Japan

Microtus nivalis   Snow vole
O   Spain to Poland to Jordan

Microtus oaxacensis   Vole
O   Mexico

Microtus ochrogaster   Prairie vole
O   Canada, United States

Microtus oeconomus   Tundra vole
O   N.Europe,Alaska,Canada,China

Microtus orcadensis   Orkney vole
O   Scotland,(Orkney Isl.)

Microtus oregoni   Oregon vole
O   British Columbia to N.W. California

Microtus pennsylvanicus   Meadow vole
O   United States, Canada,Mexico

Microtus pinetorium   Pine vole
O   E. United States, Ontario,(Canada)

Microtus quasiater   Jalapan pine vole
O   E.Mexico

Microtus roberti   Robert's vole
O   Caucasus, Asia Minor

Microtus savii   Vole
O   Italy, S. France,Spain,Portugal

Microtus sikimensis   Sikkim vole
O   Sikkim, Bhuta

Microtus socialis   Social vole
O   W.Russia,Iran, Syria

Microtus subterraneus   European pine vole
O   France to W.Russia

Microtus townsendii   Townsend's vole
O   British Columbia to N. California

Microtus transcaspicus   Transcaspian vole
O   Turkmen,Afghanistan

Microtus umbrosus   Vole
O   Mexico,(Mt.Zempoaltepec)

Microtus ungurensis   Vole
O   Transbaikalia

Microtus xanthognathus   Yellow cheeked vole
O   Alaska to Manitoba and Alberta

Myopus schisticolor   Wood lemming
O   N.Europe, N.Asia

Neofiber alleni   Round tailed muskrat
O   Florida, Georgia

Ondatra zibethicus   Muskrat
O   Canada, United States
     Z   W   P.W
                                                    P.Z

Phenacomys albipes   Pacific phenacomys
O   Oregon,California

Phenacomys intermedius   Mountain phenacomys
O   Canada, United States

Phenacomys longicaudus   Tree phenacomys
O   Oregon, California

Phenacomys silvicola   Dusky phenacomys
O   N.W.Oregon

Prometheomys schaposchikowi   Long claw mole vole
O   S.E.Russia

Synaptomys borealis   Northern bog lemming
O   Alaska to N.United States

Synaptomys cooperi   Southern bog lemming
O   S.Canada to N.E.United States

194

Ammodillus imbellis   Ammodille
O   Somalia,Ethiopa

Brachiones przewalskii   Przewalski's gerbil
O   China, Mongolia

Desmodilliscus braueri   Dwarf gerbil
O   Nigeria to Sudan

Desmodillus auricularis   Cape short tailed gerbil
O   S.Africa

Gerbillus acticola   Gerbil
O   Somalia

Gerbillus amoenus   Gerbil
O   Egypt

Gerbillus andersoni   Gerbil
O   Egypt, N.E. Sinai

Gerbillus bilensis   Gerbil
O   Ethiopia

Gerbillus bottai   Gerbil
O   Sudan,Kenya

Gerbillus calidus   Gerbil
O   Botswana

Gerbillus campestris   Rock gerbil
O   N.Africa, Sudan

Gerbillus cheesmani   Cheesman's gerbil
O   Iraq, Saudi Arabia

Gerbillus cosensi   Gerbil
O   Uganda

Gerbillus dallonii   Gerbil
O   Chad

Gerbillus dasyrus   Gerbil
O   Algeria to W.India

Gerbillus dunni   Gerbil
O   Somalia

Gerbillus eatoni   Gerbil
O   Libya

Gerbillus famulus   Black tufted gerbil
O   Saudi Arabia

Gerbillus foleyi   Gerbil
O   W.Algeria

Gerbillus gerbillus   Egyptian gerbil
O   Algeria to Jordan

Gerbillus gleadowi   Hairy footed gerbil
O   India

Gerbillus harwoodi   Gerbil
O   Kenya

Gerbillus henleyi   Pygmy gerbil
O   Sinai to Algeria

Gerbillus latastei   Gerbil
O   Morocco to Libya

Gerbillus longicaudus   Gerbil
O   Egypt

Gerbillus mackilligini   Gerbil
O   Sudan

Gerbillus macropus   Gerbil
O   Sudan

Gerbillus muriculus   Gerbil
O   Sudan

Gerbillus nancillus   Gerbil
O   Sudan

Gerbillus nanus   Naked soled gerbil
O   N.and W. Africa to Saudi Arabia

Gerbillus nigeriae   Gerbil
O   Nigeria,Mauritania

Gerbillus nigrotibialis   Gerbil
O   Angola

Gerbillus paeba   Pygmy gerbil
O   S.Africa, Botswana

Gerbillus poecilops   Aden gerbil
O   S.Arabia

Gerbillus pulcinatus   Gerbil
O   Ethiopia

Gerbillus pyramidun   Egyptian gerbil
O   Morocco to Saudi Arabia

Gerbillus rosalinda   Gerbil
O   Sudan

Gerbillus simoni   Gerbil
O   Algeria to Egypt

Gerbillus stignomyx   Gerbil
O   Sudan

Gerbillus swalius   Gerbil
O   S.W.Africa

Gerbillus vallinus   Gerbil
O   S.W.Africa

Gerbillus watersi   Gerbil
O   Sudan,Somalia

Meriones arimalius   Jird
O   C.Arabia

Meriones blackleri   Turkish jird
O   Syria, Iraq,Iran

Meriones crassus   Sundevall's jird
O   Algeria to N.W.India

Meriones hurrianae   Indian desert gerbil
O   Iran to India

Meriones libycus   Libyan jird
O   N.Africa to N.China

Meriones meridianus   Midday gerbil
O   N.Cancasus to N.China

Meriones persicus   Persian jird
O   Iran, Afghanistan

Meriones rex   King jird
O   Saudi Arabia

Meriones shawi   Shaw's jird
O   Morocco to Syria

Meriones tamariscinus   Tamarisk gerbil
O   China

Meriones tristrami   Tristam's jird
O   Israel to Iran

Meriones unguiculatus   Clawed jird
O   China

Meriones vinogradovi   Gerbil
O   Iran to N.E.Asia Minor

Meriones zarudyni   Gerbil
O   Afghanistan,Iran

Monodia mauritaniae   Gerbil
O   N.W.Mauritania

Pachyuromys duprasi   Fat tailed gerbil
O   N.Africa

Psammomys obesus   Fat sand rat
O   N.Africa to Arabia

Rhombomys opimus   Great gerbil
O   Iran to Mongolia

Sekeetamys calurus   Bushy tailed jird
O   Egypt,Syria

Tatera afra   Cape greater gerbil
O   S.Africa to Sudan

Tatera benvenuta   Gerbil
O   S.Sudan,Kenya

Tatera boehmi   Bohm's gerbil
O   Angola to Kenya

Tatera brantsi   Brant's gerbil
O   Botswana, Zimbabwe

Tatera caffer   Gerbil
O   S.Africa

Tatera cosensi   Gerbil
O   E.Tanzania

Tatera dichrura   Gerbil
O   Zaire

Tatera fallax   Gerbil
O   Uganda

Tatera flavipes   Gerbil
O   E.Sudan

Tatera fraterculus   Gerbil
O   Malawi

Tatera gambiana   Gerbil
O   W.Africa

198

Tatera guineae
O   Senegal to Ghana
Guinea gerbil

Tatera hopkinsoni
O   W.Africa
Hopkinson's gerbil

Tatera indica
O   Sri Lanka, India to Arabia
Indian gerbil

Tatera joanie
O   S.W.Africa
Gerbil

Tatera kempii
O   Gambia to Cameroon
Kemp's gerbil

Tatera leucogaster
O   Botswana,Zimbabwe
Bushveld gerbil

Tatera macropous
O   Sudan
Gerbil

Tatera minuscula
O   Ethiopia
Gerbil

Tatera nigricauda
O   Kenya
Gerbil

Tatera nigrita
O   Uganda,Zaire
Gerbil

Tatera phillipsi
O   Somalia, Kenya
Gerbil

Tatera pringlei
O   Tanzania
Black gerbil

Tatera robusta
O   Sudan
Gerbil

Tatera shoana
O   Somalia
Gerbil

Tatera taborae
O   Tanzania
Gerbil

Tatera valida
O   Angola to Malawi
Bocages gerbil

Tatera welmanni
O   Nigeria
Welman's gerbil

Taterillus congicus
O   Sudan,Zaire
Gerbil

Taterillus emini   Gerbil
O   Sudan

Taterillus gracilis   Slender gerbil
O   Senegal to Chad

Taterillus harringtoni   Gerbil
O   Zaire

Taterillus lacustris   Gerbil
O   Nigeria

Taterillus lowei   Gerbil
O   Uganda

Taterillus melanops   Gerbil
O   Kenya

Taterillus nigeriae   Nigerian gerbil
O   Nigeria

Taterillus nubilus   Gerbil
O   Kenya

Taterillus osgoodi   Gerbil
O   Kenya

Taterillus pygargus   Gerbil
O   Senegal

Taterillus tenebricus   Gerbil
O   Kenya

Beamys hindei   Long tailed pouched rat
O   Kenya

Beamys major   Long tailed pouched rat
O   Malawi, Zimbabwe

Cricetomys gambianus   African giant rat
O   Africa

Saccostomus campestris   Cape pouched mouse
O   Africa

Delanymys brooksi   Delany's swamp mouse
O   E.and C.Africa

Dendromus exoneratus   Tree mouse
O   Nigeria

Dendromus kahuziensis   Tree mouse
O   Zaire

200

Dendromus lovati   Climbing mouse
O   Ethiopia

Dendromus melanotis   Dark eared tree mouse
O   Cape to Ethiopia

Dendromus mesomelas   Mountain tree mouse
O   Mts. tropical Africa

Dendromus messorius   Tree mouse
O   Nigeria to Uganda

Dendromus mystacalis   Lesser climbing mouse
O   E.and W.Africa

Dendromus nyikae   Tree mouse
O   Zambia, Zimbabwe

Dendromus oreas   Tree mouse
O   Cameroon

Dendromus pumilo   Tree mouse
O   Sudan

Dendropionomys rousseloti   Mouse
O   Zaire

Deomys ferrugineus   Congo forest mouse
O   Central Africa

Malacothrix typicus   Mouse gerbil
O   S.Africa

Petromyscus collinus   Rock mouse
O   S.and S.W.Africa

Petromyscus monticularis   Rock mouse
O   Namibia

Prionomys batesi   Bate's tree mouse
O   C.and W. Africa

Steatomys aquilo   Fat mouse
O   Sudan

Steatomys bocagei   Fat mouse
O   Angola

Steatomys caurinus   Fat mouse
O   Nigeria to Senegal

Steatomys cupredius   Fat mouse
O   Senegal to Nigeria

Steatomys gazellae   Fat mouse
O   Sudan

Steatomys krebssi   Peter's fat mouse
O   Botswana, Zimbabwe,South Africa

Steatomys minutus   Tiny fat mouse
O   Botswana, Angola

Steatomys parvus   Fat mouse
O   E.to S. Africa

Steatomys pratensis   Fat mouse
O   Africa

Steatomys swalius   Fat mouse
O   S.W.Africa

Steatomys thomasi   Fat mouse
O   Sudan

Otomys anchietae   Swamp rat
O   Angola, Tanzania, Zimbabwe

Otomys denti   Swamp rat
O   Uganda,Zaire

Otomys irroratus   Vlei rat
O   S.E.and W.Africa

Otomys karoensis   Karro rat
O   S.Africa

Otomys laminatus   Laminate vlei rat
O   South Africa

Otomys slogetti   Swamp rat
O   South Africa

Otomys tugelensis   Vlei rat
O   S.Africa

Otomys typus   Karro rat
O   Uganda, Zaire, Rwanda

Otomys unisulcatus   Bush karro rat
O   South Africa

Parotomys brantsi   Karro rat
O   South Africa, Botswana

Parotomys littledalei   Karro rat
O   S.and S.W.Africa

Spalax ehrenbergi   Palestine mole rat
O   Syria to Libya

Spalax leucodon   Lesser mole rat
O   Hungary to Turkey

Spalax microphthalmus   Russian mole rat
O   S.Russia to Greece

Cannomys badius   Lesser bamboo rat
O   Nepal to Thailand

Rhizomys pruinosus   Hoary bamboo rat
O   Burma to Malaya

Rhizomys sinensis   Chinese bamboo rat
O   S.China,N.Burma

Rhizomys sumatrensis   Large bamboo rat
O   Burma to Sumatra

Tachyoryctes ankoliae   Mole rat
O   Uganda

Tachyoryctes annectens   Mole rat
O   Kenya

Tachyoryctes audax   Mole rat
O   Kenya

Tachyoryctes cheesmani   Mole rat
O   Ethiopia

Tachyoryctes daemon   Mole rat
O   Tanzania

Tachyoryctes macrocephalus   Mole rat
O   Ethiopia

Tachyoryctes naivashae   Mole rat
O   Kenya

Tachyoryctes pontifex   Mole rat
O   Ethiopia

Tachyoryctes rex   Mole rat
O   Kenya

Tachyoryctes ruandae   Mole rat
O   Rwanda

Tachyoryctes ruddi   Mole rat
O   Kenya

Tachyoryctes spalacinus   Mole rat
O   Kenya

Tachyoryctes splendens   Mole rat
O   Ethiopia

Tachyoryctes storeyi   Mole rat
O   Kenya

FAMILY: MURIDAE, Old World Rats & Mice

The second largest rodent family, murids are a highly varied group in diet and habitat. Their natural range includes Africa, Europe, Asia and the Australian region, mainly in tropical and subtropical areas. Through introduction by man, The genera Mus and Rattus are nearly worldwide. In many areas of the world, murids compete effectively with man for food.. They also do great crop damage and are transmitters of disease.

Acomys albigena   Spiny mouse
O   Ethiopia

Acomys cahirinus   Cairo spiny mouse
O   W.India to Africa

Acomys cinerascens   Spiny mouse
O   Sudan

Acomys demidiatus   Spiny mouse
O   Arabia,Iran

Acomys hawashensis   Spiny mouse
O   Ethiopia

Acomys hunteri   Spiny mouse
O   Sudan

Acomys hystrella   Spiny mouse
O   Uganda, Sudan

Acomys ignitus   Spiny mouse
O   Kenya

Acomys intermedius   Spiny mouse
O   Sudan

Acomys johannis   Spiny mouse
O   Nigeria

Acomys kempi   Spiny mouse
O   Kenya

Acomys louisae   Spiny mouse
O   N.Somalia

Acomys nubilis   Spiny mouse
O   Egypt

Acomys percivali   Spiny mouse
O   Uganda, Kenya, Zaire

Acomys russatus   Golden spiny mouse
O   Egypt to Arabia

Acomys spinosissimus   Spiny mouse
O   Mozambique

Acomys subspinosus   Cape spiny mouse
O   Cape Province

Acomys wilsoni   Spiny mouse
O   Kenya to Sudan

Aethomys bocagei   Bush rat
O   S.Africa

Aethomys chrysophilus   Red veld rat
O   Botswana, Zimbabwe

Aethomys kaiseri   Bush rat
O   Uganda, Sudan

Aethomys namaquensis   Rock rat
O   Botswana, Zimbabwe

Aethomys nyikae   Bush rat
O   Uganda,Zimbabwe

Aethomys paedulcus   Tree rat
O   Botswana, Zimbabwe

Aethomys stannarius   Tinfield's Rat
O   Nigeria

Aethomys thomasi   Bush rat
O   Angola

Anisomys imitator   New Guinea giant rat
O   New Guinea

Apodemus agarius   Striped field mouse
O   Europe, Asia

Apodemus flavicollis   Yellow necked mouse
O   Most of Europe, Asia Minor

Apodemus mystacinus   Broad toothed field mouse
O   Yugoslavia, Greece

205

Apodemus peninsulae O Siberia,China, Korea	Wood mouse
Apodemus speciosus O Japan	Japanese field mouse
Apodemus sylvaticus O Europe, Asia, N.W.Africa	Common field mouse
Arvicanthis lacernatus O Ethiopia	Grass rat
Arvicanthis niloticus O Egypt to W.Africa	Nile rat
Arvicanthis somalicus O Somalia,Ethiopia,Kenya	Grass rat
Bandicota bengalensis O Pakistan to Java	Lesser bandicoot rat
Bandicota indica O Sri Lanka to Java	Large bandicoot rat
Batomys dentatus O Philippines,(Luzon)	Forest rat
Batomys granti O. Philippines,(N.Luzon)	Forest rat
Carpomys melanurus O Philippines,(Luzon)	Fruit rat
Carpomys phaerus O Philippines,(Luzon)	Luzon rat
Chiromyscus chiropus O Indochina,Burma	Feas tree rat
Chiropodomys calamianensis O Philippines,(Calanianes Isl.)	Tree mouse
Chiropodomys gliroides O S.Asia	Pencil tailed tree mouse
Chiropodomys major O Borneo	Pencil tailed tree mouse
Chiropodomys muroides O Borneo	Pencil tailed tree mouse
Chiropodomys niadus O Sumatra	Tree Mouse

206

Chiropodomys pusillus   Tree mouse
O   N.Borneo

Colomys goslingi   Gosling's swamp rat
O   Africa

Conilurus albipes   White footed tree rat
O   Australia

Conilurus penicillatus   Bush tailed tree rat
O   Australia,Papua

Crateromys schadenbergi   Bush tailed cloud rat
O   Philippines,(Luzon)

Crunomys fallax   Swamp rat
O   Philippines,(Luzon)

Crunomys melanius   Swamp rat
O   Philippines (Mindanao)

Dacnomys millardi   Millard's rat
O   N.India to Laos

Dasymys incomtus   Shaggy rat
O   Africa

Dasymys nudipes   Shaggy rat
O   Angola,Zimbabwe

Diomys crumpi   Crump's mouse
O   India

Echiothrix leucura   Celebes shrew rat
O   Celebes

Eropeplus canus   Celebes gray rat
O   Celebes

Golunda ellioti   Indian bush rat
O   India,Sri Lanka

Grammomys buntingi   Thicket rat
O   Sierra Leone to Ivory Coast

Grammomys cometes   Tree rat
O   Tanzania,Kenya

Grammomys dolichurus   Tree rat
O   Africa

Grammomys gigas   Tree rat
O   Kenya

207

Haeromys humei
O   N.W.India

Humes rat

Haeromys margarettae
O   Borneo

Pygmy tree mouse

Haeromys minahassae
O   Celebes

Pygmy tree rat

Haeromys pusillus
O   Borneo

Lesser ranee mouse

Hapalomys longicaudatus
O   Malaya to Indochina

Marmoset mouse

Hybomys trivirgatus
O   Sierra Leone to Nigeria

Temminck's striped mouse

Hybomys univittatus
O   Cameroon to Uganda

Peter's striped mouse

Hyomys goliath
O   New Guinea

New Guinea giant rat

Leimacomys buttneri
O   Togo

Groove toothed forest mouse

Lemniscomys barbarus
O   Africa

Zebra mouse

Lemniscomys griselda
O   S.Africa

Single striped grass mouse

Lemniscomys lynesi
O   Sudan

Grass mouse

Lemniscomys macculus
O   Sudan to Uganda

Grass mouse

Lemniscomys striatus
O   Africa

Spotted grass mouse

Lenomys longicaudus
O   Celebes

Celebes rat

Lenomys meyeri
O   Celebes

Trefoil toothed giant rat

Leporillus apicalis
O   Australia

White tipped stick nest rat

Leporillus conditor
O   Australia

Stick nest rat

Lophuromys brevicaudus O Ethiopia	Harsh furred mouse
Lophuromys flavopunctatus O Uganda, Tanzania,Zaire	Harsh furred mouse
Lophuromys luteogaster O Zaire	Harsh furred mouse
Lophuromys major O Congo	Harsh furred mouse
Lophuromys naso O Gabon	Harsh furred mouse
Lophuromys nudicaudus O Cameroon, Nigeria	Fire bellied rat
Lophuromys rahmi O Zaire,Rwanda	Harsh furred mouse
Lophuromys sikapusi O W.Africa to Uganda	Rusty bellied rat
Lophuromys woosnami O Uganda,Zaire	Harsh furred mouse
Lorentzimys nouhuysi O New Guinea	New Guinea jumping mouse
Macruromys elegans O New Guinea	Rat
Macruromys major O N.E. New Guinea	Rat
Malacomys edwardsi O Sierra Leone to Nigeria	Long footed rat
Malacomys longipes O Ghana, Central Africa	Long footed rat
Mallomys rothschildi O New Guinea	New Guinea complex toothed rat
Mastacomys fuscus O Australia,Tasmania	Broad toothed rat
Melasmothrix naso O Celebes	Lesser shrew rat
Melomys aerosus O Ceram Isl.	Naked tailed rat

Melomys albidens   Naked tailed rat
O   New Guinea,(Lake Habbema)

Melomys cervinipes   Fawn footed melomys
O   Australia

Melomys fellowsi   Naked tailed rat
O   New Guinea

Melomys fraterculus   Naked tailed rat
O   Ceram Isl

Melomys fulgens   Naked tailed rat
O   Ceram Isl

Melomys leucogaster   Naked tailed rat
O   New Guinea

Melomys levipes   Naked tailed rat
O   New Guinea

Melomys littoralis   Grassland melomys
O   Australia

Melomys lutillus   Little melomys
O   Australia,New Guinea

Melomys moncktoni   Naked tailed rat
O   New Guinea

Melomys obiensis   Naked tailed rat
O   Obi Isl.(Indonesia)

Melomys platyops   Naked tailed rat
O   New Guinea

Melomys porculus   Naked tailed rat
O   Solomon Isl.

Melomys rubex   Naked tailed rat
O   New Guinea

Melomys rubicola   Naked tailed rat
O   Papua,(Fly river)

Melomys rufescens   Naked tailed rat
O   New Guinea

Mesembriomys gouldi   Black backed tree rat
O   Australia

Mesembriomys macrurus   Golden backed tree rat
O   Australia

Micromys minutus   Harvest mouse
O   Europe,Asia

Millardia gleadowi   Sand colored rat
O   India,Pakistan

Millardia kathleenae   Soft furred field rat
O   Burma

Millardia meltada   Soft furred field rat
O   Sri Lanka, India

Mindanaomys salomonseni   Mindanao rat
O   Philippines

Muriculus imberbis   Striped backed mouse
O   Ethiopia

Mus bellus   Mouse
O   Sudan to Angola

Mus birungensis   Mouse
O   Zaire

Mus booduga   Indian field mouse
O   India,Burma

Mus bufo   Mouse
O   Uganda,Zaire

Mus callewaerti   Mouse
O   Angola, Zaire

Mus cervicolor   Fawn colored mouse
O   Sri Lanka,India to Thailand

Mus crociduroides   Mouse
O   Sumatra

Mus deserti   Mouse
O   Angola

Mus famulus   Mouse
O   India,Burma,Thailand,China

Mus fernandoni   Mouse
O   Sri Lanka

Mus gratus   Mouse
O   Kenya,Uganda

Mus haussa   Hausa mouse
O   Niger,Nigeria

Mus imberbis   Mouse
O   Ethiopia

Mus kakhyensis   Burmese mouse
O   Viet Nam

Mus mahomet   Pygmy mouse
O   Ethiopia, Somalia

Mus mayori   Mayor's mouse
O   Sri Lanka

Mus minutoides   Mouse
O   Africa

Mus musculoides   Temminck's mouse
O   W. Africa ,Sudan

Mus musculus   **House mouse**
O   Worldwide
     Z   W   P

Mus neavei   Mouse
O   Zimbabwe,Zambia

Mus pahari   Sikkim mouse
O   Sikkim to Thailand, Java

Mus pasha   Mouse
O   Zaire

Mus platythrix   Brown spiny mouse
O   India to Burma

Mus proconodon   Mouse
O   Somalia,Ethiopia

Mus sorellus   Mouse
O   Uganda

Mus tenellus   Mouse
O   Sudan

Mus tritton   Mouse
O   Uganda, Sudan,Zimbabwe

Mus wamae   Mouse
O   Zaire

Mylomys cuninghamei   Groove toothed rat
O   Uganda

Mylomys lowei
O   W.Africa to Kenya

African groove toothed rat

Nesokia indica
O   Egypt to India

Short tailed bandicoot rat

Nesoromys ceramicus
O   Ceram Isl,(Mt.Manusela)

Ceram rat

Notomys alexis
O   C. Australia

Spinifex hopping mouse

Notomys amplus
O   Nt.

Short tailed hopping mouse

Notomys aquilo
O   Nt. Qld.

Northern hopping mouse (E*)

Notomys cervinus
O   S.C.Australia

Fawn colored hopping mouse

Notomys fuscus
O   Wa.Qld.

Dusky hopping mouse

Notomys longicaudatus
O   Australia

Long tailed hopping mouse

Notomys magalotis
O   Australia

Big eared hopping mouse

Notomys mitchelli
O   Australia

Mitchell's hopping mouse

Notomys mordax
O   Qld.

Darling Down's hopping mouse

Oenomys hypoxanthus
O   Africa

Rufous nosed rat

Papagomys armandvillei
O   Flores (East Indies)

Flores complex toothed rat

Pelomys campanae
O   Angola, Zaire

Groove toothed swamp rat

Pelomys dembeensis
O   Ethiopia

Swamp rat

Pelomys dybowskii
O   Congo

Swamp rat

Pelomys fallax
O   Uganda, Botswana, Angola

Creek rat

213

Pelomys harringtoni   Swamp rat
O   Ethiopia

Pelomys hopkinsi   Swamp rat
O   Uganda,Zaire

Pelomys isselli   Swamp rat
O   Kome Isl. (Lake Victoria)

Pelomys luluae   Swamp rat
O   Zaire

Pelomys minor   Swamp rat
O   Zimbabwe, Angola,Zaire

Pelomys rex   Swamp rat
O   Ethiopia

Phloeomys cumingi   Slender tailed cloud rat
O   Philippines

Phloeomys pallidus   Cloud rat
O   Philippines

Pithecheir melanurus   Red tree rat
O   Sumatra,Java

Pogonomelomys bruijni   Naked tailed rat
O   New Guinea

Pogonomelomys mayeri   Naked tailed rat
O   New Guinea

Pogonomelomys ruemmleri   Naked tailed rat
O   New Guinea

Pogonomelomys sevia   Naked tailed rat
O   New Guinea

Pogonomys fergussoniensis   Prehensile tailed rat
O   Fergusson Isl.

Pogonomys forbesi   Prehensile tailed rat
O   New Guinea

Pogonomys kagi   Prehensile tailed rat
O   Papua

Pogonomys lamia   Prehensile tailed rat
O   Papua

Pogonomys macrourus   Prehensile tailed rat
O   New Guinea

Pogonomys mollipilosus
O   New Guinea

Prehensile tailed rat

Pogonomys shawmayeri
O   Fergusson Isl.

Prehensile tailed rat

Pogonomys sylvestris
O   New Guinea

Prehensile tailed rat

Pogonomys vates
O   Papua

Prehensile tailed rat

Praomys albipes
O   Sudan,Ethiopia

Soft furred rat

Praomys angolensis
O   Angola,Zaire

Soft furred rat

Praomys butleri
O   Sudan

Soft furred rat

Praomys collinus
O   S.W.Africa

Soft furred rat

Praomys delectorum
O   E.Africa

Soft furred rat

Praomys denniae
O   Ugadan,Zaire,Zambia

Soft furred rat

Praomys fumatus
O   Sudan, Uganda

Soft furred rat

Praomys hartwigi
O   Cameroon (Lake Oku)

Soft furred rat

Praomys jacksoni
O   Tropical Africa

Soft furred rat

Praomys kulmei
O   Sudan

Soft furred rat

Praomys monticularis
O   S.W.Africa

Soft furred rat

Praomys morio
O   Cameroon to Uganda

Soft furred rat

Praomys natalensis
O   Africa

Soft furred rat

Praomys shortridgei
O   Botswana,Namibia

Soft furred rat

215

Praomys stella
O   Sudan,Uganda,Zaire
Soft furred rat

Praomys tullbergi
O   W.Africa, Sudan
Soft furred rat

Praomys verreauxi
O   Cape Province
Soft furred rat

Pseudomys albocinereus
O   Wa.
Ashy gray mouse

Pseudomys australis
O   Australia
Easter mouse

Pseudomys delicatulus
O   Australia
Little native mouse

Pseudomys desertor
O   Australia
Brown desert mouse

Pseudomys fieldi
O   Nt.
Alice springs mouse (E*)

Pseudomys foresti
O   Australia
Forest's mouse

Pseudomys fumeus
O   Vic.
Smokey mouse (E*)

Pseudomys gouldii
O   Australia
Gould's native mouse

Pseudomys gracilicaudatus
O   Australia
Eastern chestnut native mouse

Pseudomys hermannsburgensis
O   Australia
Pebble mound mouse

Pseudomys higginsi
O   Tasmania
Long tailed rat

Pseudomys nanus
O   Australia
Chestnut native mouse

Pseudomys novaehollandiae
O   Australia
New Holland mouse (E*)

Pseudomys occidentalis
O   Wa.
Western mouse (E*)

Pseudomys oralis
O   Qld. Nsw.
Hasting's river mouse

Pseudomys praeconis
O   Wa.
Shark bay mouse (R)

Pseudomys shortridgei
O   Australia
Blunt faced rat (E*)

Rattus adspersus
O   Celebes
Rat

Rattus adustus
O   Philippines
Rat

Rattus aeta
O   Mt. Cameroon
Beaded wood mouse

Rattus albigularis
O   Philippines
Rat

Rattus alleni
O   Fernando Po
Allen's wood mouse

Rattus alticola
O   Borneo
Mt.Spiny rat

Rattus andrewsi
O   Celebes
Rat

Rattus annandalei
O   Thailand to Sumatra
Singapore rat

Rattus arcuatus
O   Celebes,(Menkoka)
Rat

Rattus argentiventer
O   Viet Nam
Ricefield rat

Rattus baeodon
O   Borneo
Rat

Rattus bagopus
O   Philippines
Rat

Rattus baluensis
O   Borneo
Summit rat

Rattus basilanus
O   Philippines, (Basilan)
Rat

Rattus beccarii
O   N.Celebes
Rat

Rattus berdmorei
O   Thailand,Viet Nam
Gray rat

Rattus blanfordi   Blanford's rat
O   India, Sri Lanka

Rattus bowersi   Bower's rat
O   India,S.China

Rattus brahma   Rat
O   N.E.India,N.Burma

Rattus calcis   Rat
O   Philippines,(Luzon)

Rattus callitrichus   Rat
O   Philippines, Celebes

Rattus canus   Gray tree rat
O   Sumatra,Java,Borneo

Rattus celebensis   Rat
O   Celebes

Rattus chrysocomys   Rat
O   Celebes

Rattus coelestis   Rat
O   S.Celebes

Rattus coloratus   Rat
O   Philippines,(Basilan)

Rattus coxingi   Swinhoe's rat
O   S.China,Burma, Indochina

Rattus cremoriventer   Dark tailed tree rat
O   India to Borneo

Rattus cutchicus   Cutch rat
O   India

Rattus daltoni   Dalton's mouse
O   Gambia to Nigeria

Rattus dammermani   Rat
O   Celebes

Rattus defua   Defua rat
O   Sierra Leone to Ghana

Rattus doboensis   Rat
O   Aru Isl.

Rattus dollmani   Rat
O   Celebes

218

Rattus dominator   Rat
O   Celebes

Rattus edwardsi   Edward's rat
O   W.Himalayas to Sumatra

Rattus eha   Smoke bellied rat
O   Nepal to Burma

Rattus elvira   Rat
O   India

Rattus everetti   Rat
O   Philippines

Rattus exiguus   Burmese rat
O   China,Burma,Viet Nam

Rattus exulans   Pacific rat
O   S.E.Asia,New Guinea,Celebes

Rattus facetus   Rat
O   Celebes

Rattus fratrorum   Rat
O   Celebes

Rattus fulvescens   Chestnut rat
O   E.Himalayas to Malaya

Rattus fuscipes   Southern bush rat
O   Australia

Rattus gala   Rat
O   Philippines

Rattus hamatus   Rat
O   Celebes

Rattus hellwaldi   Rat
O   Celebes

Rattus hoffmani   Rat
O   Celebes

Rattus hoogerwerfi   Rat
O   Sumatra

Rattus huang   Rat
O   E.China, Viet Nam

Rattus inas   Mountain spiny rat
O   Malaya

219

Rattus infralutens   Mt. Giant rat
O   Borneo

Rattus kelleri   Rat
O   Phippines

Rattus lalolis   House rat
O   Celebes

Rattus legatus   Rat
O   Liukiu Isl.

Rattus leucophaetus   Rat
O   Philippines

Rattus leucopus   Mottle tailed Cape York rat
O   Qld.,S.New Guinea

Rattus longicaudatus   Target rat
O   Nigeria to Zaire

Rattus losea   Rat
O   China,Taiwan

Rattus luteiventris   Rat
O   Philippines,(Palawan)

Rattus lurteolus   Eastern swamp rat
O   Australia

Rattus luzonicus   Rat
O   Philippines

Rattus macleari   Rat
O   Christmas Isl.

Rattus maculipilis   Rat
O   Celebes

Rattus magnirostris   Rat
O   Philippines

Rattus manipulus   Manipus rat
O   Burma,India

Rattus marmosurus   Rat
O   N.E.Celebes

Rattus mayonicus   Rat
O   Philippines,(Luzon)

Rattus mindorensis   Rat
O   Philippines,(Mindora)

Rattus moi   Kloss rat
O   Laos, Viet Nam

Rattus montanus   Rat
O   Sri Lanka

Rattus mulleri   Muller's rat
O   Burma to Borneo

Rattus musschenbroekii   Musschenbroek's rat
O   Celebes to Malaya

Rattus nativitatus   Rat
O   Christmas Isl. ,Java

Rattus negrinus   Rat
O   Philippines

Rattus niobe   Rat
O   New Guinea

Rattus nitidus   Himalayan rat
O   Nepal to Thailand,China

Rattus niviventer   White bellied rat
O   China, Viet Nam

Rattus norvegicus   **Norway rat**
O   Worldwide
       Z   **W**   **P**

Rattus ochracieventer   Chestnut bellied spiny rat
O   Borneo

Rattus ohiensis   Ohiya rat
O   Philippines,(Cagayan Isl.)

Rattus palmarum   Rat
O   Nicobar, Andaman Isl.

Rattus pantarensis   Rat
O   Philippines,(Mindanao)

Rattus paraxanthurus   Rat
O   Celebes

Rattus pulliventer   Rat
O   Nicobar Isl.

221

Rattus punicans   Rat
O   Celebes

Rattus querceti   Rat
O   Philippines

Rattus rajah   Brown spiny rat
O   S. Burma to Borneo

Rattus rattoides   Turkestan rat
O   Afghanistan to Hainan

Rattus rattus   **House rat**
O Worldwide
     Z   W   P

Rattus rennelli   Rat
O   Rennell and Solomon Isl.

Rattus robiginosus   Rat
O   Philippines,(CagayanIsl.)

Rattus rogersi   Rat
O   S.Andaman Isl

Rattus ruber   Rat
O   New Guinea,Solomon Isl.

Rattus sabanus   Noisy rat
O   India to Borneo

Rattus salocco   Rat
O   S.E.Celebes

Rattus sladeni   Sladen's rat
O   Viet Nam,Laos,China

Rattus sordidus   Dusky field rat
O   Australia,Papua

Rattus surifer   Red spiny rat
O   S.Burma to Borneo

Rattus taerae   Rat
O   N.Celebes

Rattus tagulayensis   Rat
O   Philippines,(Mindanao)

Rattus tiomanicus   Malaysian wood rat
O   Malay Peninsula,Sumatra,Java,Borneo

Rattus tondanus   Rat
O   Australia

Rattus tunneyi   Tunney's rat
O   Australia

Rattus tyrannus   Rat
O   Philippines

Rattus verecundus   Rat
O   Papua,NewGuinea

Rattus vigoratus   Rat
O   Philippines

Rattus vulcani   Rat
O   Philippines

Rattus whiteheadi   Whitehead's rat
O   S.Thailand to Borneo

Rattus xanthurus   Rat
O   Celebes

Rattus zamboangae   Rat
O   Philippines

Rhabdomys pumilio   Four striped rat
O   Africa

Solomys ponceleti   Naked tailed rat
O   Bougainville Isl.(Solomon)

Solomys salebrosus   Naked tailed rat
O   Bougainville Isl.(Solomon)

Solomys sapientis   Naked tailed rat
O   Santa Ysabel Isl.(Solomon)

Stenocephalemys albocaudata   Narrow headed rat
O   Ethiopia

Tateomys rhinogradoides   Tate's rat
O   Celebes

Thamnomys rutilans   Shining thicket rat
O   Guinea to Zaire

Thamnomys venustus   Thicket rat
O   Uganda, Zaire

Tokudaia osimensis   Liukiu spiny rat
O   Liukiu Isl.

223

Uranomys ruddi
O   W. to E.Africa
Brush furred rat

Uromys anak
O   New Guinea
Giant naked tailed rat

Uromys caudimaculatus
O   Qld.,New Guinea
Giant naked tailed rat

Uromys imperator
O   Solomon Isl.
Giant naked tailed rat

Uromys neobritannicus
O   New Britain Isl.
Giant naked tailed rat

Uromys rex
O   Solomon Isl.
Giant naked tailed rat

Uromys salamonis
O   Solomon Isl.
Giant naked tailed rat

Vandeleuria oleracea
O   India to Thailand
Palm mouse

Vernaya fulva
O   Yunnan,N.Burma
Vernay's climbing mouse

Xenuronomys barbatus
O   New Guinea
New Guinea giant rat

Zelotomys hildegardeae
O   S.and E.Africa
Broad headed rat

Zelotomys woosnami
O   S.Africa
Rat

Zyzomys argurus
O   Australia
Common rock rat

Zyzomys pedunculatus
O   Nt.
Mac Donnell range rock rat (I)

Zyzomys woodwardi
O   Australia
Woodward's rock rat

Baiyankamys shawmayeri
O   New Guinea
Baiyanka water rat

Celaenomys silaceus
O   Philippines,(Luzon)
Gray Luzon water rat

Chrotomys whiteheadi
O   Philippines,(Luzon)
Back striped Luzon water rat

224

Crossomys moncktoni
O   Papua, New Guinea

Monckton's water rat

Hydromys chrysogaster
O   Australia,New Guinea

Beaver rat

Hydromys habbema
O   New Guinea

Beaver rat

Hydromys neobritannicus
O   New Britain Isl., New Guinea

Beaver rat

Leptomys elegans
O   Papua

Water rat

Mayermys ellermani
O   N.E.New Guinea

Shaw Mayer's mouse

Microhydromys richardsoni
O   New Guinea

Lesser water rat

Neohydromys fuscus
O   New Guinea

New Guinea water rat

Parahydromys asper
O   New Guinea

Mountain water rat

Paraleptomys rufilatus
O   New Guinea

Rat

Paraleptomys wilhemina
O   New Guinea

Rat

Pseudohydromys murinus
O   N.E.New Guinea

Rat

Pseudohydromys occidentalis
O   New Guinea

Rat

Rhynchomys soridoides
O   Philippines

Philippines shrew rat

Xeromys myoides
O   Australia

False water rat (E*)

FAMILY: GLIRIDAE, Dormice

Dormice are entirely Old World in distribution. They are small with short limbs, digits, and claws, and bushy or furred tails. They are typically nocturnal and all species hibernate. They have the ability to lose and regenerate their tails.

Dryomys nitedula
O   Europe,Asia

Forest dormouse

Eliomys melanurus
O   Syria to N.W.Arabia   Garden dormouse

Eliomys quercinus
O   Europe,Asia,N.Africa   Garden dormouse

Glirulus japonicus
O   Japan   Japanese dormouse

Glis glis
O   Europe, Asia   Fat dormouse

Muscardinus avellanarius
O   Europe, Asia Minor   Common dormouse

Myomimus personatus
O   Bulgaria,Turkey, Russia   Mouse like dormouse

Graphiurus angolensis
O   Angola,Zimbabwe   Dormouse

Graphiurus ansorgei
O   S.Angola   Dormouse

Graphiurus brockmani
O   Somalia, Kenya   Dormouse

Graphiurus crassicaudatus
O   W.Africa   Jentink's dormouse

Graphiurus hueti
O   Senegal to Angola   Huet's dormouse

Graphiurus honstoni
O   N.Zimbabwe   Dormouse

Graphiurus murinus
O   Cape to Sahara   Africana dormouse

Graphiurus ocularis
O   Cape Province   Dormouse

Graphiurus orobinus
O   Sudan   Dormouse

Graphiurus parvus
O   Africa   Dormouse

Graphiurus personatus
O   N.Uganda   Dormouse

Graphiurus platyops
O   S. Africa   Rock dormouse

Graphiurus rupicola   Dormouse
O   S.W.Africa

Graphiurus soleatus   Dormouse
O   Uganda,Tanzania

Graphiurus surdus   Dormouse
O   Cameroon, Equatorial Guinea

Graphiurus woosmani   Dormouse
O   Botswana

Platacanthomys lasiurus   Malabar spiny dormouse
O   S.India

Typhlomys cinereus   Chinese pygmy dormouse
O   Viet Nam, China

Selevinia betpakdalaensis   Betpakdala dormouse
O   Central Asia

FAMILY : ZAPODIDAE, Jumping mice

Mouse- like animal, modified for jumping;can jump as far as two meters,(6 feet). The long tail is used as a balancing organ during leaps. Gain weight in the fall and become dormant for six to eight months.

Sicista betulina   Birch mouse
O   N.Europe to Siberia

Sicista caucasica   Birch mouse
O   Russia,(Caucasus)

Sicista caudata   Far East birch mouse
O   E.Siberia

Sicista concolor   Chinese birch mouse
O   China

Sicista napaea   Birch mouse
O   Russian (Altai Mt.)

Sicista subtilis   Birch mouse
O   Czechoslovakia, Russia to Siberia

Eozapus setchuanus   Szechuan jumping mouse
O   China

Napaeozapus insignis   Woodland jumping mouse
O   Canada,United States

Zapus hudsonius   Meadow jumping mouse
O   Canada, United States, Alaska

Zapus princeps   Western jumping mouse
O   Yukon to S.W. United States

Zapus trinotatus   Pacific jumping mouse
O   British Columbia to California

FAMILY: DIPODIDAE, Jerboas

Jerboas are noted for their extreme adaptations for jumping progression (Saltation), and life in an arid environment. Their range extends from northern Africa through Arabia and Asia Minor ,and from southern Russia to Mongolia and northeastern China.
They have compact bodies, large heads, reduced forelimbs, elongated hindlimbs, and a long tufted tail. Their way of life somewhat parallels that of the kangaroo rats. They are nocturnal, many eating seeds, some depending on insects. Jerboas live in burrows which they usually plug up during the day to retain humidity.. Unlike the kangaroo rats, most species of jerboas hibernate in winter.

Alactagulus pumilio   Little earth hare
O   Iran to Mongolia

Allactaga bobrinskii   Five toed jerboa
O   Russian Turkestan

Allactaga bullata   Five toed jerboa
O   Mongolia

Allactaga elater   Five toed jerboa
O   Russia,Iran,Afghanistan

Allactaga euphractica   Euphrates jerboa
O   Asia Minor

Allactaga hotsoni   Five toed jerboa
O   Iran,Afghanistan,Pakistan

Allactaga major   Great jerboa
O   S.Russia

Allactaga severtzovi   Jerboa
O   Russia,Mongolia

Allactaga sibirica   Five toed jerboa
O   Russia, Mongolia

Allactaga tetradactyla   Four toed jerboa
O   Egypt,Libya

Allactaga williamsi   William's jerboa
O   Afghanistan

Dipus sagitta   Northern three toed jerboa
O   Iran to Manchuria

Jaculus blanfordi O Iran, Pakistan	Blanford's jerboa
Jaculus jaculus O Iraq to N. Africa to Sudan	Egyptian jerboa
Jaculus lichtensteini O Soviet C. Asia	Lechtenstein's jerboa
Jaculus orientalis O N. Africa to Israel	Egyptian jerboa
Paradipus ctenodactylus O S.W. Russia	Comb toed jerboa
Pygeretmus platyurus O Kazakh	Fat tailed jerboa
Pygeretmus shitkovi O Kazakh (Lake Balkhash)	Fat tailed jerboa
Stylodipus telum O Ukraine to Mongolia	Thick tailed three toed jerboa
Cardiocranius paradoxus O Siberia, Mongolia, China	Satunin's pygmy jerboa
Salpingotus crassicauda O N. Mongolia	Thick tailed pygmy jerboa
Salpingotus kozlovi O Mongolia, (Gobi)	Pygmy jerboa
Salpingotus thomasi O Afghanistan	Pygmy jerboa
Euchoreutes naso O China, Mongolia	Long eared jerboa

HYSTRICOMORPHS

FAMILY: HYSTRICIDAE, Old World porcupines

Old World porcupines are large rodents weighing up to 66 lbs. Some of the hair is specialized into stiff, sharp, barbless spines. The large plantigrade feet have 5 toes and smooth soles. Terrestrial animals, they dig rather extensive burrows. When threatened, they erect the quills on the head and body, rattle the tailquills, and rush towards the attacker. Large cats are their main predators.

Atherurus africanus O W. Africa	Brush tailed porcupine

Atherurus centralis O C. Africa	Brush tailed porcupine
Atherurus macrourus O China,(Szechuan) to Sumatra.	Brush tailed porcupine
Atherurus turneri O Sudan, Kenya	Brush tailed porcupine
Hystrix africaeaustralis O River Congo to Cape	Crested porcupine
Hystrix brachyurus O S. Thailand to Borneo	Malayan porcupine
Hystrix cristata O Italy, Africa	Crested porcupine
Hystrix galeata O E.Africa	Crested porcupine
Hystrix godgsoni O Nepal to S.E.China	Chinese porcupine
Hystrix indica O India to Arabia	Indian crested porcupine
Thecurus crassispinis O Borneo	Thick spined porcupine
Thecurus pumilis O Philippines	Porcupine
Trichys lipura O Borneo	Long tailed porcupine
Trichys macrotis O Malaya	Long tailed procupine

FAMILY : ERETHIZONTIDAE, New World porcupines

New World porcupines differ in several ways from the Old World family. In size, they are somewhat smaller, weighing up to 37 lb. All have stiff barbed quills which penetrate flesh more easily and are harder to remove than barbless Old World spines. Some have arboreal adaptations. New World porcupines do not burrow, but take shelter in hollow logs of overhanging rocks. They are widely distributed in forested areas. These porcupines are frequently preyed upon by the fisher, a large arboreal weasel, which turns the porcupine on its belly

Chaetomys subspinosus O Brazil	Thin spined porcupine (I)

230

| Coendou bicolor | Porcupine |
| O   Panama to Bolivia | |

| Coendou insidiosus | Porcupine |
| O   Amazonian Brazil,Surinam | |

| Coendou mexicanus | Mexican porcupine |
| O   S.Mexico to Panama | |

| Coendou pallidus | Pale porcupine |
| O   W.Indies | |

| Coendou prehensilis | Porcupine |
| O   N.South America | |

| Coendou rothschildi | Rothschild's porcupine |
| O   Panama | |

| Coendou spinosus | Porcupine |
| O   Brazil, Paraguay,Argentina | |

| Coendou vestitus | Porcupine |
| O   Venezuela, Colombia | |

| Echinoprocta rufescens | Upper Amazon porcupine |
| O   Colombia | |

Erethizon dorsatum   Porcupine
O   Alaska to N.Mexico
        Z   W   P.W
                                                                   P.Z

FAMILY :CAVIIDAE, Guinea pig, Cavy

This family includes the familiar "guinea pig" and the rabbit-like Patagonian Cavy . They
occur in most of South America

| Cavia aperea | Cavy |
| O   Brazil, Paraguay, Argentina | |

| Cavia fulgida | Cavy |
| O   E.Brazil | |

| Cavia nana | Cavy |
| O   W.Bolivia | |

| Cavia pamparum | Cavy |
| O   Argentina | |

Cavia porcellus	Guinea pig	
O Guyana to Chile		
Z	W	P.W
		P.Z

Cavia tschudii	Cavy
O Peru,S.Bolivia,Argentina	

Galea flavidens	Cavy
O Brazil	

Galea musteloides	Cavy
O Bolivia,Peru, Argentina	

Galea spixii	Cavy
O Bolivia, Brazil,Paraguay	

Kerodon rupestris	Rock cavy
O N.E. Brazil	

Microcavia australis	Cavy
O Argentina,Chile	

Microcavia niata	Cavy
O Bolivia,Peru	

Microcavia sphiptoni	Cavy
O N.W.Argentina	

Dolichotis patagonum	Patagonian cavy
O W.and S.Argentina	

Dolichotis salinicola	Mara
O Paraguay, Argentina	

FAMILY: HYDROCHOERIDAE, Capybara

Capybaras are the largest living rodents, weighing up 100 lb. They occur along borders of marshes and stream banks in the northern half of South America, East of the Andes and in Panama. They can run rapidly ,but usually seek shelter in the water where they are quite at home with good swimming and diving ability. Their chief predators are Jaguars and Anacondas.

Hydrochoerus hydrochaeris	Capybara	
O Venezuela to N. Argentina		
Z	W	P.W
		P.Z

Hydrochoerus isthmius	Panama capybara	
O E.Panama		
Z	W	P.W
		P.Z

232

FAMILY: DINOMYIDAE,Pacarana

Pacarana is a Tupi Indian word meaning "false Paca"

Dinomys branickii   Pacarana
O   Colombia to Peru

FAMILY: DASYPROCTIDAE, Agoutis and Pacas

Agoutis and pacas are relatively large with rabbit-like heads and pig-like bodies with no tail. They have short limbs but also many adaptations for running. Pacas have a unique modification of the skull and mouth, allowing them to produce a resonating rumbling sound. Agoutis and pacas inhabit tropical forests from southern Mexico through the northern half of South America. They take refuge in burrows. Both are hunted and eaten by man.

Cuniculus paca   Paca
O   Veracruz to Paraguay
     Z   W   P

Stictomys taczanowskii   Mountain paca
O   Venezuela to Ecuador
     Z   W   P

Dasyprocta aguti   Brazilian agouti
O   Guyana to Brazil
     Z   W   P

Dasyprocta albida   Agouti
O   Lesser Antilles,(St Vincent)
     Z   W   P

Dasyprocta antillensis   Agouti
O   Lesser Antilles, (St.Lucia)
     Z   W   P

Dasyprocta azarae   Agouti
O   Brazil, Paraguay
     Z   W   P

Dasyprocta coiba   Agouti
O   Panama,(Coiba Isl.)
     Z   W   P

Dasyprocta crlstata   Agouti
O   The Guianas
     Z   W   P

Dasyprocta fuliginosa   Agouti
O   N.South America
     Z   W   P

Dasyprocta kalinowski O S.E. Peru		Agouti	
Z	W		P

Dasyprocta mexicana O Mexico,(Veracruz)		Mexican agouti	
Z	W		P

Dasyprocta noblei O Guadeloupe		Agouti	
Z	W		P

Dasyprocta punctata O Mexico to Argentina		Agouti	
Z	W		P

Dasyprocta ruatanica O Honduras,(Ruatan Isl.)		Agouti	
Z	W		P

Myoprocta acouchy O Guyana to Ecuador,Brazil		Green acouchi	
Z	W		P

Myoprocta pratti   Red acouchi
O   Colombia, Peru, Brazil

FAMILY: CHINCHILLIDAE, Chinchillas, Viscachas

Chinchillids are moderately large with dense fur and long, well-furred tails. They have ever-growing cheek teeth. They occupy a variety of habitats, including the pampas (plains) and brushlands of southern South America, and the rocky slopes of Bolivia and Peru . The chinchilla is familiar to many people due to the use of its fur in making coats.

Chinchilla brevicaudata   Chinchilla (I)
O   S.Peru,S.Bolivia,Argentina

Chinchilla laniger   Chinchilla (I)
O   N. Chile

Lagidium peruanum   Mt. Viscacha
O   Peru

Lagidium viscaccia   Mt. Viscacha
O   Argentina, Chile, Bolivia,Peru

Lagidium wolffsohni   Mt. Viscacha
O   S.W.Argentina,S.Chile

Lagostomus maximus   Plains viscacha
O   Argentina,Paraguay

FAMILY: CAPROMYIDAE, Hutias, Nutria

Almost exterminated in their native habitat. The Nutria, has been introduced in Northern Hemisphere, where it has become abundant.

Capromys melanurus
O   E.Cuba

Bushy tailed hutia (R)

Capromys nana
O   Cuba.(Matanzas)

Dwarf hutia (R)

Capromys pilorides
O   Cuba

Desmarest's hutia

Capromys prehensilis
O   Cuba

Prehensile tailed hutia

Geocapromys brownii
O   Jamaica

Brown's hutia (I)

Geocapromys ingrahami
O   Bahama Isl.

Ingraham's hutia (R)

Plagiodontia eadium
O   Haiti,Dominican Republique

Cuvier's hutia (R)

Plagiodontia hylaeum
O   Haiti

Johnson's hutia

Myocastor coypus
O   Southern S.America and introduced in United States
    Z   W   P.W

Nutria

                                                         P.Z

FAMILY: OCTODONTIDAE, Octodonts

Inhabit coastal areas, foothills and the Andes Mt. up to 3500m.

Aconaemys fuscus
O   Andes of Chile,Argentina

South american rock rat

Octodon bridgesi
O   Chile

Degu

Octodon degus
O   Chile

Degu

Octodon lunatus
O   C. Chile

Degu

Octodontomys gliroides
O   Bolivia, Argentina,Chile

Boris

235

Octomys barrerae O Argentina,(Mendoza)	Viscacha
Octomys mimax O W. Argentina	Viscacha rat
Spalacopus tabanus O Chile	Coruro

FAMILY: CTENOMYIDAE, TUCO tucos

In appearance these South American burrowing rodents strongly resemble the North American pocket gophers.

Ctenomys boliviensis O Bolivia	Tuco tuco
Ctenomys brasiliensis O Brazil	Tuco tuco
Ctenomys colburni O Argentina	Tuco tuco
Ctenomys dorsalis O Paraguay	Tuco tuco
Ctenomys emilianus O Argentina	Tuco tuco
Ctenomys frater O Argentina, Bolivia	Tuco tuco
Ctenomys fulvus O W.Argentina, Chile	Tuco tuco
Ctenomys knighti O W.Argentina	Tuco tuco
Ctenomys leucodon O S.W.Bolivia, S.E.Peru	Tuco tuco
Ctenomys lewisi O S.Bolivia	Tuco tuco
Ctenomys magellanicus O S.Chile,S.Argentina	Tuco tuco
Ctenomys maulinus O Chile	Tuco tuco
Ctenomys mendocinus O Argentina	Tuco tuco

Ctenomys minutus   Tuco tuco
O   Brazil,(Mato Grosso)

Ctenomys natteri   Tuco tuco
O   Brazil

Ctenomys opimus   Tuco tuco
O   Argentina,Peru,Bolivia

Ctenomys perrensis   Tuco tuco
O   N.E.Argentina

Ctenomys peruanus   Tuco tuco
O   S.Peru

Ctenomys pontifex   Tuco tuco
O   E.Argentina

Ctenomys porteousi   Tuco tuco
O   Argentina,(Buenos Aires)

Ctenomys robustus   Tuco tuco
O   N.Chile

Ctenomys saltarius   Tuco tuco
O   N.Argentina

Ctenomys sericeus   Tuco tuco
O   S.W.Argentina

Ctenomys steinbachi   Tuco tuco
O   Bolivia

Ctenomys talarum   Tuco tuco
O   E.Argentina

Ctenomys torquatus   Tuco tuco
O   Uruguay

Ctenomys tuconax   Tuco tuco
O   N.W.Argentina

FAMILY: ABROCOMIDAE, Chinchilla rats

Some resemblance to Chinchilla, but has proportionately longer head and ears, which give a rat-like appearance.

Abrocoma bennetti   Chinchilla rat
O   Chile

Abrocoma cinerea   Chinchilla rat
O   Argentina,Chile,Peru,Bolivia

FAMILY: ECHIMYIDAE, Spiny rats

Spiny rats are small with prominent eyes and ears. The tail is quite long and can be lost readily-- an aid in escaping predators. Spiny rats occur in tropical habitats from Nicaragua to Paraguay and southeastern Brazil. Frequently they live in heavily vegetated areas near water. Entirely herbivorous, some forage in trees while others do so on the ground.

Carterodon sulcidens
O   E.Brazil
Spiny rat

Cercomys cunicularis
O   E.Brazil,Paraguay
Punare

Clyomys laticeps
O   Brazil
Spiny rat

Diplomys caniceps
O   W.Colombia,E.Ecuador
Spiny rat

Diplomys labilis
O   Panama
Darling's rat

Diplomys rufodorsalis
O   N.E.Colombia
Spiny rat

Echimys armatus
O   Guyana,Surinam,Ecuador
Guianan spiny rat

Echimys blainvillei
O   Brazil
Spiny rat

Echimys braziliensis
O   E.Brazil
Spiny rat

Echimys chrysurus
O   Guyana,Surinam
Spiny rat

Echimys dasythrix
O   S.E.Brazil
Spiny rat

Echimys grandis
O   Amazonian Brazil,Peru
Spiny rat

Echimys macrurus
O   C.Brazil
Spiny rat

Echimys nigrispinus
O   Brazil
Spiny rat

Echimys saturnus
O   Ecuador
Spiny rat

Echimys semivillosus   Spiny rat
O   Venezuela, Colombia

Echimys unicolor   Spiny rat
O   Brazil

Euryzgomatomys spinosus   Guira
O   Brazil. Paraguay

Hoplomys gymnurus   Armored rat
O   Honduras to Ecuador

Isothrix bistriatus   Toro
O   Brazil, Venezuela

Isothrix pictus   Toro
O   E.Brazil

Isothrix villosus   Toro
O   E.Peru

Lonchothrix emiliae   Spiny rat
O   Brazil

Mesomys didelphoides   Rat
O   Brazil

Mesomys hispidus   Rat
O   N.South America

Mesomys obscurus   Rat
O   Brazil

Proechimys albispinus   Spiny rat
O   Brazil,(Bahia)

Proechimys canicollis   Spiny rat
O   Colombia,Guyanas

Proechimys dimidiatus   Spiny rat
O   Brazil,(Rio de Janeiro)

Proechimys goeldii   Spiny rat
O   W.Amazonian Brazil

Proechimys guyannensis   Spiny rat
O   Venezuela to Peru,Brazil

Proechimys hendeei   Spiny rat
O   Ecuador,Peru

Proechimys iheringi   Spiny rat
O   E.Brazil

239

Proechimys longicaudatus   Spiny rat
O   Peru, Bolivia,Brazil

Proechimys myosuros   Spiny rat
O   Brazil,(Bahia)

Proechimys quadruplicatus   Spiny rat
O   Ecuador,(Lunchi Isl.)

Proechimys semispinosus   Tome's spiny rat
O   Honduras to N.Brazil

Proechimys setosus   Spiny rat
O   E.Brazil

Dactylomys bolivensis   Coro coro
O   Bolivia, Peru

Dactylomys dactylinus   Coro coro
O   N.Brazil to Ecuador

Kannabateomys amblyonyx   Tree rat
O   Brazil, Paraguay, Argentina

Thrinacodus albicauda   Rat
O   Colombia

Thrinacodus edax   Rat
O   Venezuela

FAMILY: THRYONOMYIDAE, Cane Rats

An important source of protein for many native people. In Ghana its meat sells for twice the cost of beef.

Thryonomys gregorianus   Cane rat
O   Cameroon to Ethiopia to S.Africa

Thryonomys swinderianus   Cane rat
O   Senegal to Cape Province

FAMILY: PETREOMYIDAE,Dassie rat

Somewhat squirrel-like in external appearance, but the tail is not bushy.

Petromus typicus   Dassie rat
O   S.Africa

FAMILY: BATHYERGIDAE, African Mole rats, Blesmols

The external form resembles that of other fossorial mammals.

Bathyergus janetta   Cape mole rat
O   South Africa,Namibia

Bathyergus suillus   Cape mole rat
O   South Africa

Cryptomys bolcagei   Mole rat
O   Angola

Cryptomys damarensis   Mole rat
O   Botswana

Cryptomys darlingi   Mole rat
O   Zimbabwe

Cryptomys foxi   Mole rat
O   Nigeria

Cryptomys holosericeus   Mole rat
O   Cape Province

Cryptomys hottentotus   Mole rat
O   Botswana,Tanzania

Cryptomys lechei   Mole rat
O   Zaire

Cryptomys mechowi   Mole rat
O   Zimbabwe,Angola

Cryptomys ochraceocinereus   Mole rat
O   Sudan

Cryptomys zechi   Mole rat
O   Togo

Georychus capensis   Blesmol
O   South Africa

Heliophobius argenteocinereus   Sand rat
O   Kenya to S. Africa

Heliophobius mottoulei   Sand rat
O   Zaire

Heliophobius spalax   Sand rat
O   Kenya

Heterocephalus glaber   Naked mole rat
O   E.Africa

FAMILY : CTENODACTYLIDAE,Gundis

Gundis resemble guinea pigs in external appearance. All are diurnal and are not known to hibernate or estivate.

Ctenodactylus gundi   Gundi
O   N.Africa

Ctenodactylus joleaudi   Gundi
O   Algeria

Ctenodactylus vali   Gundi
O   N.Africa

Felovia vae   West African gundi
O   W.Africa

Massoutiera harterti   Gundi
O   Algeria

Massoutiera mzabi   Latastes gundi
O   Algeria,Morocco

Massoutiera rothschildi   Gundi
O   C.Sahara

Pectinator spekei   E. African gundi
O   Ethiopia,Somalia

TOTAL RODENTIA: O	Z	W	P.W	P.Z.

ORDER: RODENTIA

ORDER: RODENTIA

ORDER: RODENTIA

_____
_____
_____
_____
_____
_____
_____
_____
_____
_____
_____
_____
_____
_____
_____
_____
_____
_____
_____
_____
_____
_____
_____

ORDER: RODENTIA

ORDER: RODENTIA

ORDER:CETACEA

This order includes the mammals most perfectly adapted to aquatic life--whales, dolphins, and porpoises.

Characteristics such as superb swimming ability, the capacity to echolocate, considerable intelligence,and well-developed social behavior have contributed to their success.

They are found in all seas of the world and in some large river systems.

Cetaceans have many structural adaptations for their completely aquatic life. The body is fusiform(cigar-shaped), lacks sebaceous glands, and is nearly hairless, which reduces drag and enables cetaceans to swim with remarkable speed and grace. Thick blubber insulates the body. The forelimbs are modified into paddle-shaped flippers with no external digits or claws. The hind limbs are vestigial and are not visible externally. The flukes (tail fins), are horizontal to the rest of the body. Propulsion is by an up and down movement of the tail flukes with the flippers serving as balancing and steering organs. Breathing is by means of one or two blowholes located on the highest point of the head. The senses of hearing and touch are more acute than that of vision. There is no sense of smell.

Some authorities divide cetaceans into two distinct orders:

ODONTOCETI: The toothed whales.

MYSTICETI: The baleen whales.

Other authorities consider these divisions to be of the suborder level.

The toothed whales Odontoceti are more diverse, more abundant and more widely distributed than are the baleen whales. Toothed whales all have teeth at some stage of life. The teeth are simple,single rooted and characteristically vary in shape and position according to species. These whales eat fish, squid and other prey which they often catch with their teeth.

The baleen whales Mysticeti have evolved a series of fringed plates, made of keratin, which hang from the roof of the mouth to form a kind of mat or sieve for filtering the plankton and other small sea life upon which these whales feed. The baleen bristles differ in number, width and length from species to species.

248

ODONTOCETI - - - Toothed Whales

FAMILY: PLATANISTIDAE,   River Dolphins

These small cetaceans have an unusually long and narrow jaw with many slender teeth. The eyes are reduced- - food and obstacles are probably detected by echolocation. River dolphins are strictly fresh- water and estuarine, found in South America and Asia, where they often inhabit murky rivers and lakes, feeding by probing muddy river bottoms.

Inia geoffrensis   Amazon dolphin
O   Rivers of N. South America

Lipotes vexillifer   Yangtze river dolphin(I)
O   Tung Ting lake,Yangtze Kiang river (China)

Platanista gangetica   Ganges river dolphin (E)
O   Rivers of India, Nepal

Pontoporia blainvillei   La Plata river dolphin
O   Atlantic coast of South America

FAMILY: DELPHINIDAE, The Oceanic Dolphins, Porpoises, Pilot Whales, Killer Whale and False Killer Whale.

The largest and most diverse group of cetaceans, the delphinids, include the classic dolphins which are smaller cetaceans with beak- like snouts and stream-lined bodies, as well as the porpoises which, although often confused with dolphins, have more rounded bodies and no projecting beak. The pilot,killer and false killer whales, also included in this group, are larger and more robust in appearance. The smallest dolphin weighs about 200 lbs., while the killer whale can weigh over 15.000lbs. and reach over 30 feet in length. Differences can also be seen among delphinids in color, shading and shape of snout.
Among the delphinids are very fast swimmers and agile leapers. They are intelligent, gregarious, and often highly social. Killer whales, which will feed on sharks, birds and other sea mammals as well as the delphinids usual fish and squid, have been observed to hunt cooperatively. The bottlenosed dolphin is widely known as a performer in sea parks and on television, where highly developed innate learning behaviors are used in "shows". They have a sophisticated echolocation system and are very vocal.

Feresa attenuata   Pygmy killer whale
O   Senegal.Japan, Hawaii waters

Globicephala macrorhynchus   Indian pilot whale
O   Tropical

Globicephala melaena   Pilot whale
O   Atlantic,Pacific

Orcaella brevirostris   Irrawaddy dolphin
O   Indian, South Pacific ocean

Orcinus orca   Killer whale
O   World wide

249

THE OCEANIC DOLPHINS,PORPOISES,PILOT WHALES,
KILLER WHALE AND FALSE KILLER WHALE

Peponocephala electra
O   Tropical oceans   Broad beaked dolphin

Pseudorca crassidens
O   World wide   False killer whale

Lissodelphis borealis
O   N. Pacific   Northern right whale dolphin

Lissodelpins peroni
O   S.Pacific   Southern right whale dolphin

Cephalorhynchus commersoni
O   S.America (Atlantic)   Commerson's dolphin

Cephalorhynchus eutropia
O   Chile coast   Black dolphin

Cephalorhynchus heavisidei
O   Cape of Good Hope   Gray dolphin

Cephalorhynchus hectori
O   New Zealand   White headed dolphin

Delphinus bairdii
O   Gulf California, Mexico   Saddle back dolphin

Delphinus delphis
O   World wide   Common dolphin

Grampus griseus
O   World wide   Risso's dolphin

Lagenodelphis hosei   Bornean dolphin
O   Tropical waters (Pacific,Atlantic,Indian Oceans)

Lagenorhynchus acutus
O   N.Atlantic   White sided dolphin

Lagenorhynchus albirostris
O   N.Atlantic   White beaked dolphin

Lagenorhynchus australis
O   S.South America   Black chinned dolphin

Lagenorhynchus cruciger   Hourglass dolphin
O   Temperate waters of the Southern Hemisphere

Lagenorhynchus obliquidens   Pacific white sided dolphin
O   N.Pacific

Lagenorhynchus obscurus
O   S.Atlantic, S.Pacific

Dusky dolphin

Stenella coeruleoalbas
O   Tropical and Temperate waters

Striped dolphin

Stenella dubia
O   Tropical waters

Spotted dolphin

Stenella longirostris
O   Tropical waters

Long beaked dolphin

Stenella roseiventris
O   Banda Sea,(Indonesia)

Spinner dolphin

Tursiops truncatus
O   World wide

Bottle nosed dolphin

Sotalia fluviatilis
O   Amazon river

Amazonian white dolphin

Sotalia guianensis
O   N.E.Coast of South America

Guiana white dolphin

Sousa borneensis
O   South China sea

Borneo white dolphin

Sousa chinensis
O   Coast of S.China

Chinese white dolphin

Sousa lentiginosa
O   S.Africa to India

Speckled dolphin

Sousa plumbea
O   Coastal N.Indian ocean

Plumbeous dolphin

Sousa teuszi
O   Coastal of W.Africa

Cameroon dolphin

Steno bredanensis
O   Tropical and Temperate waters

Rough toothed dolphin

Neophocaena phocaenoides
O   From Pakistan to Korea

Black finless porpoise

Phocoena dioptrica
O   S.Atlantic

Spectacled porpoise

Phocoena phocoena
O   Water of N. Hemisphere

Harbor porpoise

251

Phocoena sinus
O   Gulf of California

Pacific harbour porpoise (V)

Phocoena spinipinnis
O   Peru to Tierra Del Fuego

Burmeister's porpoise

Phocoenoides dalli
O   N.Pacific

Dall's porpoise

FAMILY: MONODONTIDAE,Beluga or White Whale and Narwhal

These medium- sized cetaceans inhabit the northern seas. In summer the belugas move far up some large rivers in Siberia and Alaska, and the St. Lawrence in Canada. They are quite vocal, having in their repetoire a musical trill which earned them the nickname "sea canary " Belugas have about 36 teeth and feed on fish and squid. The narwhal is noted for the long, straight, forward- directed tusk of the male, one of its only two teeth. The function of the tusk is unknown.

Delphinapterus leucas
O   Arctic and northern water

White whale

Monodon monoceros
O   North polar seas

Narwhal

FAMILY: PHYSETERIDAE, Sperm Whales

This family contains the giant sperm whale ,60 ft. long ,100.000 lbs; the pygmy sperm whale 13ft. long,700lbs; and dwarf sperm whale 8 ft.long,500 lbs.which inhabit all oceans. The head of the giant sperm whale is huge and very distinctive with a squared snout and narrow underslung jaw. In males, which are very much larger than females, the head may account for over one third of the body length, The head contains great quantities of oil (spermaceti), which has been prized by man for years. Giant sperm whales feed mainly on deep water squids as well as bony fishes, sharks and skates. They are quite social, assembling in groups sometimes as large as 1000 individuals, Some adult males are solitary. Males migrate far north to the edge of the pack ice in summer, but females stay in temperate and tropic waters.Much less is known of the pygmy and dwarf sperm whales, although both are probably fairly abundant. Smaller than the giant sperm, the pygmy sperm is somewhat shark-like in appearance and does not seem to be gregarious. The dwarf sperm is even smaller, but can be mistaken easily for the pygmy.

Kogia breviceps
O   World wide

Pygmy sperm whale

Kogia simus
O   World wide

Dwarf sperm whale

Physeter catodon
O   World wide.

Sperm whale (E*)

FAMILY: ZIPHIIDAE, Beaked Whales

Beaked whales are medium sized cetaceans, with slender bodies and noticeably beaked snouts. The dentition is reduced in all but one species (Tasmacetus shepherdi). In some species a mandibular tooth is visible, protruding outside the mouth. Widely distributed, beaked whales are nevertheless poorly known and some species have never been seen alive. Some are deep divers that can stay submerged for long periods (over an hour). They forage primarily on squid, but also on deep sea fishes.

Berardius arnouxi   Arnoux's beaked whale
O   S. Pacific

Berardius bairdii   Baird's beaked whale
O   N. Pacific

Hyperoodon ampullatus   Bottle nosed whale (V)
O   Waters of N. Hemisphere

Hyperoodon planifrons   South bottle nosed whale
O   Waters of S.Hemisphere

Indopacetus pacificus   Indo pacific whale
O   Qld.

Mesoplodon bidens   Sowerby's beaked whale
O   N.Atlantic

Mesoplodon bowdoini   Deep crested whale
O   New Zealand, Tasmania

Mesoplodon carlhubbi   Arch beaked whale
O   N.Pacific

Mesoplodon densirostris   Blainvilles beaked whale
O   World wide

Mesoplodon europaeus   Antillean beaked whale
O   W. N. Atlantic

Mesoplodon ginkgodens   Ginko toothed whale
O   N. Pacific

Mesoplodon grayi   Gray beaked whale
O   S.Pacific

Mesoplodon hectori   Deep socketed whale
O   Pacific and Indian Oceans

Mesoplodon layardi   Strap toothed whale
O   S.Africa, S.Australia,S.America

Mesoplodon mirus   True's beaked whale
O   Atlantic

253

Mesoplodon stejnegeri   Stejneger's beaked whale
O   N.Pacific

Tasmacetus shepherdi   Tasmanian beaked whale
O   S.Pacific

Ziphius cavirostris   Goose beaked whale
O   World wide

MYSTICETI - - - Baleen Whales

FAMILY:ESCHRICHTIDAE,   Gray Whale

The gray whale is about 50 feet long, relatively slender bodied and has no dorsal fin. It is black or slaty gray with many white spots and blotches, some of which are white barnacles. They live in Arctic seas in the summer, feeding largely on botom dwelling crustaceans. In autumn, they migrate southward and winter along coastlines - - the western Pacific group along the coast of Korea; the eastern Pacific group along the coast of Baja California. The round trip migration is from 6.500 to 14.500 miles- - the longest known mammalian migration.Gray whale populations declined greatly between 1850 and 1925 from whaling, but they are now protected and have increased in recent years. They are still threatened by pollution and disturbance by humans at the whale's breeding gounds.

Eschrichtius robustus   Gray whale (E*)
O   N.Pacific

FAMILY:BALAENOPTERIDAE,   Rorquals

This family varies in size from some rather small whales to the Blue Whale;the largest mammal that has ever lived. It reaches a length of 100 feet and may weigh 300.000 lbs. The skin of the throat and chest of rorquals, is marked by furrows which increase the capacity of the mouth when opened. Rorquals live in all oceans feeding in cold waters.They migrate south in the Northern Hemisphere and north in the Southern Hemisphere, during the winter. Adults do not feed during this period but live off stored blubber. Whaling has greatly reduced the numbers of Fin whales, Humpbacks, and Blue whales. The numbers of blue whales have been so drastically reduced that they may not be able to recover.

Balaenoptera acutorostrata   Lesser rorqual
O   World wide

Balaenoptera borealis   Sei whale (E*)
O   World wide

Balaenoptera edeni   Brydes whale
O   Tropical coastal waters

Balaenoptera musculus   Blue whale (E)
O   World wide

Balaenoptera physalus   Common rorqual (V)
O   World wide

254

Megaptera novaeangliae   Humpback whale (E)
O   World wide

FAMILY: BALAENIDAE, Right Whales

These large whales reach a length of about 60 ft. and have huge heads- - over 1/3 their total length. The flippers are short and rounded, the dorsal fin usually absent. They live in most oceans, but are absent from tropical and south polar seas. They are most commonly found near coastlines or near pack ice and are not highly migratory. Right whales were killed in great numbers by whalers and are rare and protected today.

Balaena glacialis   Right whale
O   All Oceans and Seas

Balaena mysticetus   Bowhead whale (E)
O   Arctic

Caperea marginata   Pygmy right whale
O   Southern oceans

TOTAL CETACEA: O   Z   W   P.W.   P.Z

ORDER: CETACEA

255

ORDER: CETACEA

ORDER: CETACEA

ORDER: CARNIVORA

The order Carnivora consists of seven recent families, which inhabit all the continents and adjoining islands throughout the world.
They range in size from the least weasel to kodiak bear and are divided into two main groups (super- families);

The Canoidea or Dog group includes, Canidae, Ursidae, Procyonidae,and Mustelidae.

The Feloidea or Cat group contains Viverridae, Hyaenidae and Felidae.

Differentiation into these two groupings is on the basis of:

1. fossil evidence of evolutionary relationships

2. similarities of internal skull characteristics and

3. the presence or absence of certain accessory sex glands.

Carnivores are primarily hunters with a good sense of smell. While not all members are strictly carnivorous,the majority include meat in their diet, and the most striking adaptations reflect this predacious way of life. Carnivores are universally thought of in terms of "tooth and claw" . Indeed, it is in theses two areas where the highest degree of spreadable digits on each foot, usually armed with strong curved claws which function to seize and hold prey. In some forms the claws are retractile and thereby protected when not in use.

The teeth are differentiated into incisors, canines and cheek teeth (premolars and molars). The canines function as stabbing teeth. The cheek teeth vary in number and function- - bears have 6 upper and 7 lower cheek teeth some well adapted for crushing- - cats have only 3 upper and lower cheek teeth none of which is adapted for cruhing. The 4 th. upper premolar and 1 st. lower molar are called carnassials and are used for shearing- - working very much like a pair of scissors. The carnassials, although present throughout the order, range from poor development in the bears and procyonids, to the highest degree of efficiency in the cats.

Locomotion is either digitigrade "toe-walking", plantigrade "sole-walking", or somewhere in between. The majority have vibrissae 'whiskers" on each cheek, used as tactile (sensory) organs and most highly developed in the more predacious forms.

Scent glands are present in most carnivores, and are used variously for territorial marking, social recognition, and defense. Off- Spring are altricial (born in a helpless condition) and require concentrated and extended parental care

SUPER FAMILY: CANOIDEA

FAMILY: CANIDAE, Dogs,Wolves,Coyotes,Jackals,Foxes

Canids are familiar animals and occur naturally almost throughout the world. They are characterized by a long muzzle, large ears, and long slender legs with blunt non-retractile claws. They are digitigrade (walk on toes). The canine teeth are long and strong, the molars retain crushing surfaces for dealing with food other than meat. Canids are not as specialized for a strictly carnivorous diet as are cats.

Canids are generally cursorial(running animals), their adaptations enabling them to run swiftly to overtake their prey ,which they detect mainly by a remarkable sense of smell.

Some canids (i.e. African hunting dogs, wolves),live and hunt in packs (groups), with quite complicated social structures. Others(i.e. foxes) are solitary, living and hunting alone or in pairs. Canids are opportunistic, feeding on vertebrates, mollusks,carrion and many types of plant material. Coyotes in western United States are reported to feed heavily on melon crops, also on wild fruits such as cactus fruits and juniper berries.

Alopex lagopus   Arctic fox
O   Circumpolar holarctic
         Z   W   P.W
                                                          P.Z

Atelocynus microtis   Small eared fox
O   N.South America
         Z   W   P.W
                                                          P.Z

Canis adustus   Side striped jackal
O   Africa
         Z   W   P.W
                                                          P.Z

Canis aureus   Golden jackal
O   Asia,Europe,Africa
         Z   W   P.W
                                                          P.Z

Canis familiaris   **Domestic dog**
O   World wide in association with people

Canis familiaris dingo   Dingo
O   Australia
         Z   W   P.W
                                                          P.Z

Canis latrans   Coyote
O   Alaska to Costa Rica
         Z   W   P.W
                                                          P.Z

Canis lupus   Gray wolf (V)
O   Holarctic
         Z   W   P.W
                                                          P.Z

Canis mesomelas   Black backed jackal
O   Sudan to South Africa
         Z   W   P.W
                                                          P.Z

259

Canis rufus   Red wolf (E)
O   Texas,Louisiana,Arkansas
    Z   W   P.W
                                     P.Z

Canis simensis   Simenian jackal
O   Ethiopia
    Z   W   P.W
                                     P.Z

Cerdocyon thous   Crab eating fox
O   South America
    Z   W   P.W
                                     P.Z

Chrysocyon brachyurus   Maned wolf (V)
O   Brazil, Bolivia,Paraguay,Argentina
    Z   W   P.W
                                     P.Z

Cuon alpinus   Dhole (V)
O   Russia, India,S.E.Asia
    Z   W   P.W
                                     P.Z

Dusicyon culpaeus   Culpeo fox
O   Andes from Ecuador southwards
    Z   W   P.W
                                     P.Z

Dusicyon griseus   Chilla
O   Argentina, Chile
    Z   W   P.W
                                     P.Z

Dusicyon gymnocercus   Azaras fox
O   South America
    Z   W   P.W
                                     P.Z

Dusicyon sechurae   Sechura fox
O   Peru,Ecuador
    Z   W   P.W
                                     P.Z

Dusicyon vetulus   Hoary fox
O   Brazil
    Z   W   P.W
                                     P.Z

Fennecus zerda   Fennec
O   N. Africa to Sudan
    Z   W   P.W
                                     P.Z

Lycaon pictus   Africa hunting dog (E*)
O   Africa
    Z   W   P.W
                                     P.Z

Nyctereutes procyonoides   Raccoon dog
O   Russia, China, Japan
    Z   W   P.W
                                     P.Z

Otocyon megalotis   Bat eared fox
O   E.and S.Africa
     Z   W   P.W
                               P.Z

Speothos venaticus   Bush dog (V)
O   Panama to N. South America
     Z   W   P.W
                               P.Z

Urocyon cinereoargenteus   Gray fox
O   Canada to Venezuela
     Z   W   P.W
                               P.Z

Urocyon littoralis   Island gray fox
O   Islands off the California coast
     Z   W   P.W
                               P.Z

Vulpes bengalensis   Bengal fox
O   India, Nepal, Pakistan
     Z   W   P.W
                               P.Z

Vulpes cana   Blanford's fox
O   Iran ,Pakistan,Afghanistan
     Z   W   P.W
                               P.Z

Vulpes chama   Cape fox
O   S.Africa
     Z   W   P.W
                               P.Z

Vulpes corsac   Corsac fox
O   C.Asia
     Z   W   P.W
                               PZ

Vulpes macrotis   Kit fox
O   S.W. United States to Mexico
     Z   W   P.W
                               P.Z

Vulpes pallida   Pale fox
O   Africa
     Z   W   P.W
                               P.Z

Vulpes ruppelli   Sand fox
O   N. Africa to Afghanistan
     Z   W   P.W
                               P.Z

Vulpes velox   Swift fox
O   Canada, United States
     Z   W   P.W
                               P.Z

Vulpes vulpes   Red fox
O   Holarctic
     Z   W   P.W
                               P.Z

261

Bears are large, heavily built, big-headed animals,with short, powerful limbs and short tails. The limbs have 5 digits with strong, recurved claws, which can be used to kill prey, tear apart logs, grasp a fish, or pull down berry laden branches. Bears are not designed for speed, they walk plantigrade (flat-footed like man), and run in a loping fashion. The teeth are different from those of other carnivores, but well adapted to bear's omnivorous diet. The molars are greatly enlarged, with "wrinkled" surfaces adapted for crushing. They can cope with many foods including insects, vertebrates, berries, grass, and carrion. Polar bears are more exclusively carnivorous feeding mainly on seals.

Bears inhabit the northern hemisphere and northern South America. Most bears live in the temperate zones. One exeption is the polar bear, which lives in the arctic. Many of their habitats have cold winters, during which bears sleep in caves or other protected areas. They do not truly hibernate;the temperature and metabolic rate do not drop greatly. There is considerable debate as to the taxonomic status of the giant panda. Some classify it as an ursid, others as a procyonid, still others put it in a separate family of its own.

Ailuropoda melanoleuca   Giant panda (R)
O   China
     Z   W   P.W
                                 P.Z

Helarctos malaynus   Sun bear
O   S.E.Asia
     Z   W   P.W
                                 P.Z

Melursus ursinus   Sloth bear (I)
O   India, Sri Lanka,Nepal,Bangladesh
     Z   W   P.W
                                 P.Z

Selenarctos thibetanus   Asiatic black bear
O   Asia
     Z   W   P.W
                                 P.Z

Tremarctos ornatus   Spectacled bear
O   South America
     Z   W   P.W
                                 P.Z

Ursus arctos   Brown bear
O   Holarctic
     Z   W   P.W
                                 P.Z

Ursus americanus   American black bear
O   Alaska to Mexico
     Z   W   P.W
                                 P.Z

Ursus maritimus   Polar bear
O   Circumpolar
     Z   W   P.W
                                 P.Z

Raccoons, coatimundis, ringtails, kinkajous, and lesser pandas belong to this varied family which occupy tropical and temperate regions of the Americas and eastern Asia.

Procyonids are small, generally with a short broad face and ringed tail. Many have facial markings. The limbs are fairly long; each foot has 5 toes with either semi-retractile or non-retractile claws. Procyonids are usually good climbers. One member, the kinkajou, has a prehensile tail. In some species,(for example the raccoon), the forepaws have considerable dexterity and are used in food handling. The teeth are adapted to an omnivorous diet which is characteristic of most procyonids- - the premolars are small and sharp, the molard broad and low crowned, the shearing action of carnassials of most procyonids is nearly lost.

Ailurus fulgens   Red panda
O   China,Nepal,Burma
      Z   W   P.W
                                             P.Z

Bassaricyon gabbii   Olingo
O   Nicaragua to N.South America
      Z   W   P.W
                                             P.Z

Bassariscus astutus   Ringtail
O   S.W.United States to Mexico
      Z   W   P.W
                                             P.Z

Bassariscus sumichrasti   Cacomistle
O   Mexico to Panama
      Z   W   P.W
                                             P.Z

Nasua narica   Coati
O   Arizona,New Mexico to Argentina
      Z   W   P.W
                                             P.Z

Nasua nasua   Ring tailed coati
O   C. and S. America
      Z   W   P.W
                                             P.Z

Nasua nelsoni   Cozumel island coati
O   Yucatan (Cozumel Isl.)
      Z   W   P.W
                                             P.Z

Nasuella olivacea   Mountain coati
O   N.South America
      Z   W   P.W
                                             P.Z

Potos flavus   Kinkajou
O   Mexico to Brazil
      Z   W   P.W
                                             P.Z

Procyon cancrivorus   Crab eating raccoon
O   Costa Rica to Uruguay
      Z   W   P.W
                                             P.Z

263

Procyon insularis			Tres Marias raccoon	
O Mexico,(Tres Marias Isl.)				
Z		W		P.W
				P.Z
Procyon lotor			Raccoon	
O Canada to Panama				
Z		W		P.W
				P.Z
Procyon maynardi			Bahaman raccoon	
O Bahamas isl.				
Z		W		P.W
				P.Z
Procyon minor			Guadeloupe raccoon	
O Guadeloupe Isl.				
Z		W		P.W
				P.Z
Procyon pygmaeus			Cozumel Isl. raccoon	
O Mexico, (Cozumel Isl.)				

FAMILY: MUSTELIDAE,Weasels, Badgers, Skunks, Otters

These small, long bodied carnivores have short limbs; claws never fully retractile; plantigrade or digitigrade feet; flattened faces; well- developed anal scent glands, and a long tail. The teeth vary in number and function. The carnassials in many species have well developed shearing edges, and in others, such as the sea otter, they are adapted for crushing. The pelage is either uniform in color, striped, or spotted. In some species it is quite beautiful and of great value to the fur trade.
Mustelids are very successful, occupying all terrestrial habitats from arctic tundra to tropical rain forests. Some are aquatic and live in rivers, lakes or the ocean. The natural distribution is nearly world wide. They are absent only from Madagascar, Australia, Antarctica,and most oceanic islands. Mustelids are basically carnivorous, hunting mainly by scent, although their senses of sight and hearing are also well developed.

Eira barbara			Tayra	
O Mexico to Argentina				
Z		W		P
Galictis cuja			Little grison	
O C.and S. South America				
Z		W		P
Galictis vittata			Grison	
O Mexico to Brazil				
Z		W		P
Gulo gulo			Wolverine	
O Holarctic				
Z		W		P.W
				P.Z
Ictonyx striatus			Zorilla	
O Africa				
Z		W		P

Lyncodon patagonicus		Patagonian weasel	
O Argentina and Chile			
Z	W		P
Martes americana		American marten	
O North America			
Z	W		P.W
			P.Z
Martes flavigula		Yellow throated marten	
O Asia			
Z	W		P
Martes foina		Stone marten	
O Palearctic			
Z	W		P
Martes gwatkinsi		Nilgiri	
O India			
Z	W		P
Martes martes		Pine marten	
O Europe, Asia			
Z	W		P.W
			P.Z
Martes melampus		Japanese marten	
O Japan,Korea			
Z	W		P
Martes pennanti		Fisher	
O North America			
Z	W		P.W
			P.Z
Martes zibellina		Sable marten	
O Palearctic			
Z	W		P
Mustela africana		Tropical weasel	
O Brazil,Ecuador,Peru			
Z	W		P
Mustela altaica		Alpine weasel	
O Russia, Tibet, China			
Z	W		P
Mustela erminea		Ermine	
O Holarctic			
Z	W		P.W
			P.Z
Mustela frenata		Long tailed weasel	
O S.Canada to Bolivia			
Z	W		P.W
			P.Z
			Yellow bellied weasel
Mustela kathiah			
O Himalayas,S.China			
Z	W		P

265

Mustela lutreola			European mink
O	Europe,Asia		
	Z	W	P.W
			P.Z
Mustela nigripes			Black footed ferret (E)
O	Canada, United States		
	Z	W	P.W
			P.Z
Mustela nivalis			Weasel
O	Palaerarctic		
	Z	W	P.W
			P.Z
Mustela nudipes			Malaysian weasel
O	Malaysia,Borneo,Sumatra		
	Z	W	P
Mustela putorius			Polecat
O	Holarctic		
	Z	W	P.W
			P.Z

Mustela putorius furo   **Domestic ferret**
O

Mustela rixosa			Least weasel
O	North America		
	Z	W	P.W
			P.Z
Mustela sibirica			Siberian weasel
O	Russia to Japan		
	Z	W	P.W
			P.Z
Mustela vison			American mink
O	North America		
	Z	W	P
Poecilictis libyca			Libyan striped weasel
O	Africa		
	Z	W	P
Poecilogale albinucha			White naped weasel
O	Uganda to South Africa		
	Z	W	P
Vormela peregusna			Marbled polecat
O	Rumania to Asia Minor		
	Z	W	P
Mellivora capensis			Ratel
O	Africa, S.Asia		
	Z	W	P
Arctonyx collaris			Hog badger
O	China to Sumatra		
	Z	W	P

Meles meles		European badger	
O Palaearctic			
Z	W		P.W
			P.Z
Melogale moschata		Chinese ferret badger	
O S.China to Indochina			
Z	W		P
Melogale orientalis		Javan ferret badger	
O Java, Borneo			
Z	W		P
Melogale personata		Burmese ferret badger	
O Nepal to Java			
Z	W		P
Mydaus javanensis		Stink badger	
O Borneo, Java,Sumatra			
Z	W		P
Suillotaxus marchei		Palawan badger	
O Philippines			
Z	W		P
Taxidea taxus		American badger	
O Canada to Mexico			
Z	W		P.W
			P.Z
Conepatus chinga		Skunk	
O N.South America			
Z	W		P
Conepatus humboldti		Patagonia skunk	
O S. America			
Z	W		P
Conepatus leuconotus		Hog nose skunk	
O Arizona,New Mexico,Texas to Mexico			
Z	W		P.W
			P.Z
Conepatus mesoleucus		Hog nosed skunk	
O Colorado to Nicaragua			
Z	W		P.W
			P.Z
Conepatus semistriatus		Amazonian skunk	
O Mexico to Brazil			
Z	W		P.
Mephitis mephitis		Striped skunk	
O Canada to Mexico			
Z	W		P.W
			P.Z
Mephitis macroura		Hooded skunk	
O Arizona to Nicaragua			
Z	W		P.W
			P.Z
Spilogale gracilis		Western spotted skunk	
O Washington to Colorado			
Z	W		P.W
			P.Z

267

Species	O (Range)	Z	W	P
Spilogale putorius	British Columbia to Costa Rica		W	P.W / P.Z
	Spotted skunk			
Spilogale pygmaea	Mexico		W	P
	Pygmy spotted skunk			
Aonyx capensis	Africa		W	P
	Cape clawless otter			
Aonyx cinerea	Asia		W	P
	Oriental small clawed otter			
Aonyx congica	Congo basin		W	P
	Congo otter			
Aonyx microdon	Nigeria, Cameroon		W	P
	Cameroon otter (R)			
Aonyx philippsi	Rwanda, Uganda, Burundi		W	P
	Philip's otter			
Enhydra lutris	Coast of Alaska to California, Japan		W	P.W / P.Z
	Sea otter			
Lutra canadensis	North America		W	P.W / P.Z
	Canadian river otter			
Lutra felina	Coastal Peru, Chile		W	P
	Sea cat (E)			
Lutra longicaudis	Mexico to South America		W	P.W / P.Z
	Otter			
Lutra prococax	Chile, Argentina		W	P
	Otter (I)			
Lutra lutra	Europe, Asia		W	P.W / P.Z
	European otter (V)			
Lutra maculicollis	Africa		W	P
	Spotted necked otter			
Lutra perspicillata	Iraq to Sumatra		W	P
	Smooth Indian otter			

268

Lutra sumatrana   Hairy nosed otter
O S.Asia, Sumatra, Borneo
  Z   W   P
Pteronura brasiliensis   Giant otter (E)
O South America
  Z   W   P

SUPER FAMILY: FELOIDEA

FAMILY: VIVERRIDAE,Civets, Genets, Mongooses

Viverrids are small, short- legged, long tailed carnivores inhabiting forested areas or dense brush of the Old World. The mongoose, which feeds mainly on snakes, has been introduced into areas of the New World. This has usually been disastrous to native fauna. Most viverrids are carnivorous, hunting prey in trees as well as on the ground. Their feet are either plantigrade or digitigrade, their claws partly retractile. The carnasials have good shearing ability. Their senses of sight, smell, and hearing, are well-developed. Like mustelids , most have scent glands in the anal region which secrete a strong smelling fluid as a defense measure. One member, the binturong, has a prehensile tail and spends most of its time in the trees. Viverrids are the largest group of carnivores in terms of number of species and yet very little is known about their habits.

Genetta abyssinica   Abyssinia genet
O Ethiopia
  Z   W   P
Genetta angolense   Angolan genet
O Angola, Zambia,Congo
  Z   W   P
Genetta genetta   European genet
O Africa, Europe
  Z   W   P
Genetta lehmanni   Lehmann's genet
O Africa
  Z   W   P
Genetta pardina   Pardine genet
O Gambia to Cameroon
  Z   W   P
Genetta servalina   Servaline genet
O Africa
  Z   W   P
Genetta tigrina   Blotched genet
O South Africa
  Z   W   P
Genetta victoriae   Giant genet
O Zaire,Uganda
  Z   W   P
Genetta villiersi   False genet
O Senegal to Lake Chad
  Z   W   P
Osbornictis piscivora   Fishing genet
O Zaire
  Z   W   P

Poiana richardsoni			Oyan	
O	W. Africa to Zaire			
	Z	W		P
Prionodon linsang			Banded linsang	
O	S.E.Asia			
	Z	W		P
Prionodon pardicolor			Spotted linsang (E*)	
O	Nepal to Indochina			
	Z	W		P
Viverra civetta			Africa civet	
O	Africa			
	Z	W		P
Viverra megaspila			Large spotted civet (E)	
O	S.Asia			
	Z	W		P
Viverra tangalunga			Oriental civet	
O	S.E.Asia, Philippines			
	Z	W		P
Viverra zibetha			Large Indian civet	
O	S.E.Asia, China, Nepal			
	Z	W		P
Viverricula indica			Small Indian civet	
O	Asia			
	Z	W		P
Arctictis binturong			Binturong	
O	S.E.Asia			
	Z	W		P
Arctogalidia trivirgata			Small toothed palm civet	
O	S.E.Asia			
	Z	W		P
Macrogalidia musschenbroeki			Brown palm civet (R)	
O	Celebes			
	Z	W		P
Nandinia binotata			Two spotted palm civet	
O	Africa			
	Z	W		P
Paguma larvata			Masked palm civet	
O	Asia			
	Z	W		P
Paradoxurus hermaphroditus			Common palm civet	
O	India to China, S.E.Asia			
	Z	W		P
Paradoxurus jerdoni			Jerdon's palm civet	
O	India			
	Z	W		P
Paradoxurus zeylonensis			Golden palm civet	
O	Sri Lanka			
	Z	W		P
Chrotogale owstoni			Owston's palm civet	
O	Viet Nam,Laos			
	Z	W		P

270

Cynogale bennetti		Otter civet	
O S.E.Asia			
Z	W		P
Eupleres goudoti		Falanouc (V)	
O Madagascar			
Z	W		P
Eupleres major		Falanouc	
O Madagascar			
Z	W		P
Fossa fossa		Fanaluka	
O Madagascar			
Z	W		P
Hemigalus derbyanus		Banded palm civet	
O S.E.Asia			
Z	W		P
Hemigalus hosei		Hose's palm civet	
O Borneo			
Z	W		P
Galidia elegans		Ring tailed mongoose	
O Madagascar			
Z	W		P
Galidictis fasciata		Mongoose	
O Madagascar			
Z	W		P
Galidictis ornata		Broad striped mongoose	
O Madagascar			
Z	W		P
Galidictis striata		Broad striped mongoose	
O Madagascar			
Z	W		P
Mungotictis lineata		Narrow striped mongoose	
O Madagascar			
Z	W		P
Mungotictis substriatus		Narrow striped mongoose	
O Madagascar			
Z	W		P
Salanoia olivacea		Salano	
O Madagascar			
Z	W		P
Salanoia unicolor		Brown tailed mongoose	
O Madagascar			
Z	W		P
Atilax paludinosus		Marsh mongoose	
O Africa			
Z	W		P
Bdeogale crassicauda		Bushy tailed mongoose	
O Kenya to Mozambique			
Z	W		p
Bdeogale jacksoni		Jackson's mongoose	
O Kenya, Uganda			
Z	W		P

271

Bdeogale nigripes   Black legged mongoose
O   Nigeria to Angola
         Z   W   P
Crossarchus alexandri   Congo kusimanse
O   Zaire, Uganda
         Z   W   P
Crossarchus ansorgei   Angolan kusimanse
O   Angola
         Z   W   P
Crossarchus obscurus   Kusimanse
O   Sierra Leone to Cameroon
         Z   W   P
Cynictis penicillata   Yellow mongoose
O   S.Africa
         Z   W   P
Helogale dybowskii   Mongoose
O   C. Africa
         Z   W   P
Helogale hirtula   Dwarf mongoose
O   E.Africa
         Z   W   P

Helogale macmillani   Mongoose
O   Africa
         Z   W   P
Helogale parvula   Dwarf mongoose
O   Africa
         Z   W   P
Helogale percivali   Percival's dwarf mongoose
O   E.Africa
         Z   W   P
Helogale vetula   Mongoose
O   E.Africa
         Z   W   P
Helogale victorina   Lake Victoria dwarf mongoose
O   E.Africa
         Z   W   P
Herpestes auropunctatus   Small Indian mongoose
O   Iraq to S.Asia
         Z   W   P
Herpestes brachyurus   Short tailed mongoose
O   S.Asia, Philippines
         Z   W   P
Herpestes dentifer   Mongoose
O   E.Africa
         Z   W   P
Herpestes edwardsi   Indian gray mongoose
O   Arabia to India
         Z   W   P
Herpestes fuscus   Indian brown mongoose
O   India, Sri Lanka
         Z   W   P

Herpestes granti			Mongoose
O E.Africa			
Z	W		P
Herpestes hosei			Hose's mongoose
O Borneo			
Z	W		P
Herpestes ichneumon			Egyptian mongoose
O Spain, Africa			
Z	W		P
Herpestes javanicus			Javan mongoose
O S.Asia, Java			
Z	W		P
Herpestes naso			Cameroon's mongoose
O W.Africa			
Z	W		P
Herpestes ochracea			Ochraceous mongoose
O C.Africa			
Z	W		P
Herpestes pulverulentus			Cape gray mongoose
O S. Africa			
Z	W		P
Herpestes sanguineus			Slender mongoose
O Africa			
Z	W		P
Herpestes semitorquatus			Collared mongoose
O Borneo, Sumatra			
Z	W		P
Herpestes smithi			Ruddy mongoose
O India,Sri Lanka			
Z	W		P
Herpestes urva			Crab eating mongoose
O Nepal, China, S.E.Asia			
Z	W		P
Herpestes vitticollis			Striped necked mongoose
O S.Asia, India, Sri Lanka			
Z	W		P
Ichneumia albicauda			White tailed mongoose
O Africa, Arabia			
Z	W		P
Liberiictis kuhni			Kuhn's kusimanse
O Liberia			
Z	W		P
Mungos gambianus			Gambian mongoose
O Gambia to Nigeria			
Z	W		P
Mungos mungo			Banded mongoose
O Africa			
Z	W		P
Paracynictis selousi			Selous meerkat
O S.Africa			
Z	W		P

Rhynchogale melleri O S.Africa		Meller's mongoose	
Z	W		P
Suricata suricatta O S.Africa		Slender tailed meerkat	
Z	W		P
Cryptorocta ferox O Madagascar		Fossa (V)	
Z	W		P

FAMILY : HYAENIDAE, Hyenas, Aardwolf

A large head, and powerful forequarters, with smaller, sloping, comparatively weak hindquarters are characteristic of hyaenids. The hyenas have a heavy build with strongly built skulls and powerful dentition, which is used to crush large bones and well developed carnassials. The claws are blunt and non retractile, the feet digitigrade. In contrast, the aardwolf is more lightly built with a delicate skull and smaller teeth , well adapted for an insectivorous diet. The aardwolf is striped and has a mane of long hair from neck to rump. Its anal scent glands are used as a protective device. When pursued, it emits a musk fluid. Hyaenids inhabit plains, brushlands , and open forests of Africa, southwestern Asia, and parts of India. They are mainly nocturnal.Though well adapted to feed on carrion, it has recently been discovered that spotted hyenas also kill prey, and lions will feed on a carcass of a hyena kill. Hyaenids tend to be very vocal emitting a number of calls and cries. The spotted hyena has a cry that sounds like laughter therefore earning it name "laughing hyena".

Crocuta crocuta O Africa		Spotted hyena	
Z	W		P.W P.Z
Hyaena brunnea O Africa		Brown hyena	
Z	W		P.W P.Z
Hyaena hyaena O Africa,Asia		Striped hyena (E)	
Z	W		P.W P.Z
Proteles cristatus O S.Africa		Aardwolf	
Z	W		P.W P.Z

FAMILY: FELIDAE, Cats, Lion, Tigers, Cheetah, Leopards

Although cats as a group are quite uniform in appearance, from the house cat to the tiger, there is no general agreement as to their taxonomy. They occur naturally throughout the world, with the exceptions of Antarctica, Australia,Madagascar, and some isolated oceanic islands.

Cats are characterized by a muscular, deep chested body with a shortened face, large eyes, and well developed vibrissae. Many are spotted or striped which helps conceal them from their prey. The number of teeth is reduced. The carnassials are well developed and specialized to enhance shearing ability, and adaptation for their completely carnivorous diet. Cats are digitigrade, the forelimbs are strong; the sharp claws are fully retractile, with the exception of the cheetah whose claws are partly retractile. Cats are very efficient hunters. They typically stalk their prey then capture it with a brief burst of speed. The cheetah is an exception. It can run at speeds approaching 70 mph. for over 600 yards.

Cats rely mainly on vision for finding food which consists of fish, mollusks, birds, and small and large mammals, depending on the size and habitat of the cat.

Felis aurata   African golden cat
O   Senegal to Kenya
       Z   W   P.W
                                                 P.Z

Felis badia   Bay cat (R)
O   Borneo
       Z   W   P.W
                                                 P.Z

Felis bengalensis   Leopard cat (E*)
O   Asia
       Z   W   P.W
                                                 P.Z

Felis bieti   Chinese desert cat
O   China
       Z   W   P.W
                                                 P.Z

Felis caracal   Caracal
O   Africa,Asia
       Z   W   P.W
                                                 P.Z

Felis catus   **Domestic cat**
O   World wide

Felis chaus   Jungle cat
O   Egypt, Asia
       Z   W   P.W
                                                 P.Z

Felis colocolo   Pampas cat
O   S. America
       Z   W   P.W
                                                 P.Z

Felis concolor   Puma (E*)
O   Yukon to Argentina
       Z   W   P.W
                                                 P.Z

Felis geoffroyi   Geoffroy's ocelot
O   S.America
     Z   W   P.W
                                     P.Z

Felis guigna   Kodkod
O   Chile, Argentine
     Z   W   P.W
                                       P.Z

Felis iriomotensis   Iriomote cat (E)
O   Iriomote Isl, (Taiwan)
     Z   W   P.W
                                       P.Z

Felis jacobita   Mountain cat (R)
O   Peru, Chile,Argentina
     Z   W   P.W
                                       P.Z

Felis libyca   African wild cat
O   Africa, Asia
     Z   W   P.W
                                       P.Z

Felis lynx   Lynx
O   Holarctic
     Z   W   P.W
                                       P.Z

Felis manul   Pallas's cat
O   Iran to Tibet,China
     Z   W   P.W
                                       P.Z

Felis margarita   Sand cat
O   Africa, Asia
     Z   W   P.W
                                       P.Z

Felis marmorata   Marbled cat (E*)
O   Asia
     Z   W   P.W
                                       P.Z

Felis nigripes   Black footed cat (E*)
O   S.Africa
     Z   W   P.W
                                       P.Z

Felis pardalis   Ocelot (V)
O   Arizona and Texas to Argentina
     Z   W   P.W
                                       P.Z

Felis pardina   Spanish lynx (E)
O   Spain, Portugal
     Z   W   P.W
                                       P.Z

Felis planiceps   Flat headed cat (E*)
O   S.E.Asia
     Z   W   P.W
                                       P.Z

Felis rubiginosus   Rusty spotted cat
O   India, Sri Lanka
      Z   W   P.W
                                                                         P.Z

Felis rufus   Bobcat
O   Canada to Mexico
      Z   W   P.W
                                                                         P.Z

Felis serval   Serval (E*)
O   Africa
      Z   W   P.W
                                                                         P.Z

Felis silvestris   European wild cat
O   Europe, Asia
      Z   W   P.W
                                                                         P.Z

Felis temmincki   Asiatic golden cat (E*)
O   Asia
      Z   W   P.W
                                                                         P.Z

Felis tigrinus   Tiger ocelot (V)
O   Costa Rica to Argentina
      Z   W   P.W
                                                                         P.Z

Felis viverrinus   Fishing cat
O   Pakistan to S.E.Asia
      Z   W   P.W
                                                                         P.Z

Felis wiedii   Margay cat (V)
O   Texas to Argentina
      Z   W   P.W
                                                                         P.Z

Felis yagouaroundi   Jaguarondi (E*)
O   Arizona, Texas to Argentina
      Z   W   P.W
                                                                         P.Z

Panthera leo   Lion
O   Africa, Asia
      Z   W   P.W

                                                                         P.Z

Panthera nebulosa   Clouded leopard (V)
O   Asia
      Z   W   P.W
                                                                         P.Z

Panthera onca   Jaguar (V)
O   S.W.United States to Argentina
      Z   W   P.W

                                                                         P.Z

Panthera pardus   Leopard (V)
O   Asia,Africa
      Z   W   P.W

                                                        P.Z

Panthera tigris   Tiger (E)
O   Asia
      Z   W   P.W

                                                        P.Z

Panthera uncia   Snow leopard (E)
O   Highlands of central Asia
      Z   W   P.W
                                                        P.Z

Acinonyx jubatus   Cheetah (V)
O   Africa, Asia
      Z   W   P.W
                                                         P.Z

TOTAL: CARNIVORA: O	Z	W	P.W.	P.Z.

_____
_____
_____
_____
_____
_____
_____
_____
_____
_____
_____
_____
_____
_____
_____
_____
_____
_____
_____
_____
_____
_____
_____
_____

ORDER:PINNIPEDIA

Pinnipedia, the seals, sea lions, fur seals and walrus, are found along cold currents in all the oceans of the world.

They probably evolved from early Carnivore types, and in some classifications the Pinnipedia (aquatic carnivores), and the Fissipedia (land carnivores), are regarded as suborders of the Carnivora.

Pinnipeds of today are entirely carnivorous in their diet.

They spend much of their lives in the water, but are not considered to be completely aquatic, as all must return to land or ice floes to give birth.

Spectacular swimmers and divers, these mammals exhibit many remarkable adaptations for an aquatic life.

Members of this sub-family (sea lions and fur seals), are easily identified because they have small, visible external ears and powerful forelimbs. They are quite agile on land as the hind flippers can be brought beneath the body to allow them to walk, or even run or climb, using all 4 flippers. The front flippers are used for locomotion in the water. They generally live in warmer waters than either the walrus or "true "seals. Their diet consists largely of fish and squid, as well as other mollusks and crustaceans.. They are gregarious and vocal; often found in large aggregations. The males, which are much larger than the females, establish territories along beaches or rocky shore areas during breeding season, which they may defend more or less vigorously against other males.

The walrus is found in relatively shallow Arctic waters near shorelines of Atlantic and Pacific oceans.They remain close to and spend a good deal of time on the pack ice. Walrus feed along the sea bottom, taking clams, crabs, worms,etc...,using their large "moustache" of short, sensitive vibrissae to find food, and their mobile lips to suck flesh from shells. This extremely large, sparsely haired pinniped, can move its hind flippers forward, (like the eared seals), to achieve four-footed progression on land, although movement is slow and ponderous.

Both sexes have long tusks formed by development of the upper canine teeth, with longest tusks found on mature males.

Arctocephalus australis   South America fur seal
O   W.Coasts of South America
     Z   W   P

Arctocephalus forsteri   New Zealand fur seal
O   Southern Australia,New Zealand
     Z   W   P

Arctocephalus galapagoensis   Galapago fur seal
O   Galapagos Isl.
     Z   W   P

Arctocephalus gazella   Kerguelen fur seal
O   Antarctic
     Z   W   P

Arctocephalus philippii   Juan Fernandez fur seal (V)
O   Chile,(Juan Fernandez Isl.)
     Z   W   P

Arctocephalus pusillus   South Africa fur seal
O   South Africa, Australia
     Z   W   P

Arctocephalus townsendi   Guadalupe fur seal (V)
O   Baja California
     Z   W   P

Arctocephalus tropicalis   Amsterdam island fur seal
O   Amsterdam and St. Paul Isl.
     Z   W   P

Callorhinus ursinus   Alaska fur seal
O   North Pacific
     Z   W   P.W
                                              P.Z

Eumetopias jubatus   Steller sea lion
O   W.Coast of North America
     Z   W   P

Neophoca cinerea   Australian sea lion
O   Australia
     Z   W   P

Otaria flavescens   South America sea lion
O   South America
     Z   W   P
Phocarctos hookeri   New Zealand sea lion
O   New Zealand
     Z   W   P
Zalophus californianus   California sea lion
O   Japan, N. America, Galapagos
     Z   W   P.W
                                                   P.Z
Odobenus rosmarus   Walrus
O   Arctic ocean
     Z   W   P.W
                                                   P.Z

FAMILY: PHOCIDAE, True seals or Earless seals

The most abundant pinnipeds, the phocids, are also a more varied family in behavior and habitat. They have no visible external ears, smaller foreflippers, and more flexible spines than Otariids, and are generally considered to be more highly specialized for an aquatic life. The hind flippers, which are used to propel the animal through the water, are useless on land, where the forelimbs are used to drag the body along slowly and awkwardly.

Cystophora cristata   Hooded seal
O   Arctic and Atlantic waters
     Z   W   P
Erignathus barbatus   Bearded seal
O   Arctic Ocean and adjoining seas
     Z   W   P
Halichoerus grypus   Gray seal
O   N. Atlantic
     Z   W   P
Phoca caspica   Caspian seal
O   Caspian sea
     Z   W   P
Phoca fasciata   Ribbon seal
O   Japan to Alaska
     Z   W   P
Phoca groenlandica   Harp seal
O   Hudson Bay , Gulf of St. Lawrence to Siberia
     Z   W   P
Phoca hispida   Ringed seal
O   Arctic ocean
     Z   W   P
Phoca kurilensis   Kuril seal (I)
O   Japan
     Z   W   P
Phoca largha   Larga seal
O   Siberia, Yukon to China
     Z   W   P

286

Phoca sibirica		Baikal seal		
O Lake Baikal				
Z	W		P	
Phoca vitulina		Habor seal		
O North Pacific and Atlantic				
Z	W		P.W	
			P.Z	
Hydrurga leptonyx		Leopard seal		
O Antarctica				
Z	W		P	
Leptonychotes weddelli		Weddell seal		
O Antarctica				
Z	W		P	
Lobodon carcinophagus		Crabeater seal		
O Antarctica				
Z	W		P	
Mirouga angustirostris		Northern elephant seal		
O Alaska to Baja California				
Z	W		P.W	
			P.Z	
Mirouga leonina		Southern elephant seal		
O Antarctica				
Z	W		P	
Monachus monachus		Mediterranean monk seal (E)		
O Mediterranean sea				
Z	W		P	
Monachus schauinslandi		Hawaiian monk seal (E)		
O Hawaiian Isl.				
Z	W		P.W	
			P.Z	
Monachus tropicalis		Caribbean monk seal		
O West Indies				
Z	W		P	
Ommatophoca rossii		Ross seal		
Antarctica				

TOTAL PINNIPEDIA: O	Z	W	P.W.	P.Z

287

ORDER:TUBULIDENTATA

The Aardvark, whose name means "earth pig", is a large (up to 5 ft. long plus tail 180 lbs), thick skinned, sparsely haired, burrowing animal with an elongate skull.
It has no incisors or canines; the slender columnar cheek teeth have no enamel but are surrounded by cement. All its digits have strong, flattened, blunt claws.. An acute sense of smell enables the Aardvark to hunt ants and termites. The powerful forelimb are used for tearing apart termite nests and for burrowing, while the hindlimbs are used to push the loose soil out of the burrow. Slow runners, their best defense is burrowing which they do quite rapidly.
They are native to most of Africa south of the Sahara, restricted in distribution only by the availability of food.
Normally one young is born at the beginning of the second rainy season (October-November) and remains in the burrow for about two weeks, at which time it starts accompanying its mother on nightly feeding trips.
It can obtain its own food at approximately 6 months of age.

FAMILY: ORYCTEROPODIDAE, Aardvark

Orycteropus afer   Aardvark
O   Africa

ORDER:PROBOSCIDEA

Great size , trunk and tusks easily distinguish the Elephant, _the largest ,living, land mammal._

The body is supported on long, massive, columnar legs, that give a powerful rather than speedy stride. Each foot has 5 digits, with five hoof- like nails on fore feet in both species and four nails on hind feet of Asiastic, and three nails on hind feet of African. Elephants walk on the tips of the digits (digitigrade), but a heel pad of dense connective tissue braces the toes, and largely transmits the weight of the animal, from the legs, to the ground. The pads spread when weight is placed on them, contract when weight is removed, which aids in getting the foot out of swampy ground.

The trunk is an elongation of the nose and upper lip, with nostrils at the tip. It retains the basic functions of breathing and smelling, but is also used for transferring food to the mouth and is capable of both powerful and delicate manipulation. Elephants can no more drink through the nose than any other mammal; rather the water is drawn into the trunk than blown into the mouth. The skull is massive-- necessary for attachment of muscles to support and use the trunk and tusks. Its bones are thick, but contain many air spaces which reduces the weight.

Elephants are herbivorous, eating a variety of leaves, vines, grasses, fruits, etc... To break down this fibrous matter their dentition is unique. In addition to the tusks, which are modified incisors, they have complex grinding cheek teeth, which are replaced horizontally from back to front, rather than vertically. Normally there is only one functional cheek tooth at a time in each section of jaw. In its lifetime an elephant will have have six (6) "sets" of cheek teeth for a total of twenty-four (24).

Sexually mature at 8-12 years, elephants reach full growth at about 25 years. Gestation is from 20- 22 months, with one young weighing approximately 200lbs., being born. Social animals, they travel in herds of various sizes. African elephant herds are led by an old cow. The herd consists of females and young. When young bulls reach maturity, they are forced to leave the herd.

FAMILY : ELEPHANTIDAE, Elephants

Loxodonta africana   African elephant (V)
O   Africa
     Z   W   P.W

                                            P.Z

Elephas maximus   Indian elephant (E)
O   Asia
     Z   W   P.W

                                            P.Z

TOTAL ELEPHANTIDAE: O	Z	W	P.W.	P.Z

ORDER: HYRACOIDEA

Hyraxes have a superficial resemblance to rodents in size and appearance, but based on technical details of the skeleton and certain other characteristics, a common ancestry for hyraxes, elephants and sirenians is postulated.

Roughly rabbit size, they have short skulls, specialized incisors, a broad diastema (gap) and grinding cheek teeth. The forefoot has 4 digits, hind foot 3, and the soles of the feet have naked pads for traction, with special muscular arrangement that permits retraction of the middle of the sole, to create a suction cup effect for climbing. The suction cup effect is enganced by a moist grandular secretion.

Agile climbers and jumpers, hyraxes occupy many habitats - - forests, scrub, grassland, lava flows - - some living in rocky ledges, other species living in trees. All are more or less gregarious. They are found only in Africa and parts of the Middle East. They are primarily herbivorous but also eat insects and grubs.

FAMILY: PROCAVIIDAE, Hyraxes

Dendrohyrax arboreus   Tree hyrax
O   Zaire, Kenya to South Africa

Z	W		P.W
			P.Z

Dendrohyrax dorsalis   Beecroft's hyrax
O   Gambia to Uganda

Z	W		PW
			P.Z

Dendrohyrax validus   Eastern tree hyrax
O   Tanzania

Z	W		P.W
			P.Z

Heterohyrax antineae   Hyrax
O   Algeria

Z	W		P.W
			P.Z

Heterohyrax brucei   Yellow spotted hyrax
O   Egypt to South Africa

Z	W		P.W
			P.Z

Heterohyrax chapini   Chapin's hyrax
O   Zaire

Z	W		P.W
			P.Z

Procavia capensis   Rock hyrax
O   Sinai Peninsula, Africa

Z	W		P.W
			P.Z

TOTAL HYRACOIDEA: O   Z   W   P.W   P.Z

ORDER: SIRENIA

Sirenians are large, nearly hairless, totally aquatic mammals with flipper- like forelimbs a vestigial pelvis, and a horizontal fluke tail. They are slow- moving grazers who feed while submerged on aquatic plants in coastal waters, large rivers, and some lakes. The teeth are unusual- the style of replacement is similar to that of elephants.

FAMILY: DUGONGIDAE,Dugong

Dugong is found in the tropical coastal areas of the Old World, confined to marine waters. While there are many differences between the two families of sirenians in teeth and skeleton, the most obvious distinction is in the shape of the tail. Dugons have a notched fluke like that of whales. The usual size of dugongs is 8- 11ft. with an average weight around 350 lbs. They have keen hearing and taste but poor vision. Breeding apparently takes place throughout the year, with a single young born after an 11 month gestation.
One recently extinct member of this family, the Steller's Sea Cow, was the only sirenian adapted to cold waters. An inhabitant of the Bering Sea, it was discovered in 1769 and exterminated within 27 years. It was more than twice as long as the dugong with an estimated weight of near 9.000lbs.

Dugong dugon   Dugong (V)
O   Pacific, Indian ocean
        Z   W   P

FAMILY: TRICHECHIDAE, Manatees

Larger than dugongs, the manatees may reach lengths of 15 ft.and approach 1500lb. although average sizes are considerably smaller 800 lbs. The tail fluke in these species is more or less evenly rounded, described by some as spoon-shaped . Each of the three species is found in geographically separate areas, but all seem to prefer the turbid waters of marine bays and sluggish rivers. One species is found along the coast and rivers of South Eastern United States, West Indies, and Northern South America
Manatees probably reach breeding maturity at 3-4 years. Gestation is between 5 and 6 months, with one or two young born underwater at any time of year and cared for by both parents; The young nurse for 18 months but probably staying with the mother for 2 years.

Trichecus inunguis   Amazonian manatee (V)
O   Amazon river and its tributaries
        Z   W   P
Trichecus manatus   American manatee (V)
O   S.E North America to N.E.Sou th America
        Z   W   P
Trichecus senegalensis   African manatee (V)
O   Tropical West Africa
        Z   W   P

TOTAL SIRENIA: O	Z	W	P.W	P.Z

ORDER: PERISSODACTYLA

Odd-toed hoofed ungulates (horse, tapir, rhino).

The diagnostic characteristic of this order is that the weight is borne on the central digit of the foot (digit # 3).

The hind foot has either 3 toes (2,3,& 4) or 1 toe (3rd.). The front foot may have 4 digits, but the weight is still carried on digit # 3. Bones of the feet are elongated and strengthened, but reduced in number.

Perissodactyls walk on the tip of the terminal phalanges, which are flattened and triangular. Each digit is encased in a hoof. These are herbivorous animals with massive grinding teeth to break down the cellulose.

Perissodactyls are on the decline. The entire order is endangered.

Members of this family have only one functional digit (the 3rd.digit).They are swift runners, with all senses well developed. Complex high crowned cheek teeth and chisel-like incisors characterize these grazers. Equidae developed originally in North America but for some unknown reason, horses disappeared from the New World before historic times.

Equus asinus   African wild ass,donkey, burro (E)
O   Once found from Morocco to Somalia, Arabian Peninsula
     Z   W   P.W

                                                  P.Z
Equus burchelli   Common zebra
O   Ethiopia to Angola and South Africa
     Z   W   P.W

                                                    P.Z

Equus caballus   **Horse**
O   Worldwide

Equus grevyi   Grevys zebra (E)
O   Kenya, Ethiopia, Somalia
     Z   W   P.W

                                                    P.Z
Equus hemionus   Asiatic wild ass (V)
O   Syria to India
     Z   W   P.W

                                                    P.Z
Equus przewalskii   Przewalski's wild horse (E)
O   Mongolia, China
     Z   W   P.W

                                                    P.Z
Equus zebra   Mountain zebra (V)
O   S. Africa
     Z   W   P.W

                                                    P.Z

FAMILY: TAPIRIDAE, Tapirs

Tapird are the most generalized of living Perissodactyls. They are heavy bodied (about 650 lbs.) with short limbs. The fore foot has 4 toes, the hind foot 3 toes, all with hooves. The nose and upper lip form a short proboscis. Tapirs mostly inhabit tropical areas, usually near water. Three species are found in South America, one in South East Asia; however, like equids, tapirs developed originally in North America and migrated into their current distribution.
Soft browse,largely succulent plants and fruit make up the diet of tapirs. They breed throughout the year with a single young born after 390-400 days. The young of all species look the same- - like "brown striped watermelons with legs".

Tapirus bairdi   Baird's tapir (V)
O   Mexico to Ecuador

Z   W   P.W

                                        P.Z

Tapirus indicus   Malayan tapir (E)
O   Malay peninsula, Burma,Thailand,Sumatra

Z   W   P.W

                                        P.Z

Tapirus pinchaque   Mountain tapir (V)
O   Andes from N.W. Venezuela to Peru

Z   W   P.W

                                        P.Z

Tapirus terrestris   South American tapir (E*)
O   Colombia,Venezuela to Argentina and Brazil

Z   W   P.W
                                        P.Z

FAMILY: RHINOCEROTIDAE, Rhinoceros

These "armored tanks" of the animal world are surviving members of an illustrious group of mammals with a complex fossil record. (one Asian form, Baluchitherium, was 18 ft. high at the shoulder- - the largest known land mammal). Living rhinos are large, stout bodied herbivores with fairly short legs. There are 3 toes on each foot. Unique to rhinos are the horns (one or two), which are supported by the nasal bones. Unlike the true horns found on cattle and allies, the horns of rhinos do not have any bony core, or direct attachment to the bone.Strong structures which grow throughout the animal's life,they are often worn dow by being rubbed against rocks or tree trunks. Unlike true horns, they will regenerate if broken.

Rhinos inhabit tropical and subtropical areas of Asia and Africa,living in many habitats. Gestation is approximately 17-18 months (510-570 days). The single young is active soon after birth and remains with the mother until another young is born, an event which does not happen for several years.

Ceratotherium simum   White rhinoceros (E)
O   Africa

Z   W   P.W

                                        P.Z

Diceros bicornis   Black rhinoceros (V)
O   Africa

Z   W   P.W

                                        P.Z

Dicerorhinus sumatrensis   Sumatran rhinoceros (E)
O   S.Asia,Sumatra,Borneo

Z   W   P.W

                                        P.Z

Rhinoceros sondaicus   Javan rhinoceros (E)
O   Sumatra,Javan,Malay peninsula
    Z   W   P.W

                                         P.Z

Rhinoceros unicornis   Indian rhinoceros (E)
O   Nepal, N. Pakistan and India
    Z   W   P.W

                                         P.Z

TOTAL PERISSODACTYLA: O	Z	W	P.W.	P.Z.

297

ORDER: ARTIODACTYLA

Pigs, hippoptamuses,camels, deer, cows, sheep, goats and antelope are all members of this order- - the most successful order of hooved animals.

They occur naturally throughout the world, with the exception of Australia, New Zealand, Antarctica, and some isolated islands.

Like perissodactyls,artiodactyls walk on the tips of their toes (unguligrade), which are encased in hooves.

The bulk of the weight of artiodactyls, however, is borne evenly on the third and fourth digits. The first digit is absent, the second and fifth digits are either reduced, vestigial or absent. The bones of limbs and feet are elongated. Some are fused for extra strength. Uniquely shaped,the ankle bone (astragalus) permits great extension of the limb and therefore great sprinting ability, but it limits the motion to forward and backward, completely restricting lateral movememt. These adaptations make artiodactyls superb runners.

Their sight, hearing and sense of smell are also well developed.

The principal diet of most artiodactyls consists of grasses and woody material of relatively low nutritional value. Artiodactyls are adapted to obtain the maximum nutritional value from this food. The cheek teeth are specialized for grinding materials high in cellulose. The stomach is quite complex to aid in digestion.

Most authorities divide artiodactyls into two suborders- -Suina: which includes pigs, peccaries, and hippopotamus, and Ruminantia; consisting of camels, giraffes, deer,antelope, sheep, goats and cattle.. The 3 families that follow are members of the suborder Suina.

They have molars with hill-shaped cusps (bunodont), which are not as specialized for an herbivorous diet as are the molars of ruminants..Their canines are tusk-like (also the incisors in hippos). They retain four complete and separate digits and have a non-ruminating (no cud chewing) stomach of 2 or 3 chambers.

Pigs are fairly large animals (weighing up to 600 lbs.), with long heads, short necks, stocky bodies,and short limbs. The skin is thick and sparsely haired. The large canines are ever-growing. The upper canines form tusks that curve upwards and backwords. The mobile snout has a disk-like pad of cartilage, and is used in turning up surface soil in search of food. Suids are native only to Old World, but have been introduced into the New World by man. They occupy mainly forested or brushy areas. They swim well and are fond of mud. Most are gregarious, some assembling in groups of up to 50 in number., Omnivorous animals, their diet includes many plants, roots, carrion, small rodents, and snakes. When attacked, suids can use their tusks which make formidable weapons.

Young piglets are born in a less precocial state than most artiodactyls. They are striped in all but two genera (domestic pig Sus and Babyrousa).

Babyrousa babyrusa   Babirusa (V)
O   Celebes
        Z   W   P.W
                                                          P.Z

Hylochoerus meinertzhageni   Giant forest hog
O   Liberia to Ethiopia and Tanzania
        Z   W   P.W
                                                          P.Z

Phacochoerus aethiopicus   Wart hog
O   Africa
        Z   W   P.W

                                                          P.Z

Potamochoerus porcus   Bush pig
O   Africa, Madagascar
        Z   W   P.W

                                                          P.Z

Sus barbatus   Bearded pig
O   Sumatra, Borneo, Philippines
        Z   W   P.W

                                                          P.Z

Sus salvanius   Pygmy hog (E)
O   S. Asia
        Z   W   P.W

                                                          P.Z

Sus scrofa   Wild boar
O   Europe, Asia, Africa
        Z   W   P.W
                                                          P.Z

Sus scrofa domestica   **Domestic hog**

Sus verrucosus   Javan pig
O   Java, Celebes, Philippines
     Z   W   P.W

                                                        P.Z

FAMILY: TAYASSUIDAE, Peccaries

Peccaries superficially resemble suids, but there are many important differences.. Peccaries are much smaller, weighing about 66lbs. The limb structure is more specialized, the lateral digits (2and 5) are smaller. The forefoot has 4 digits, the hind foot, 3 digits. The snout ,like that of pigs, is mobile and is used in digging for food. Peccaries are omnivorous, but they tend to rely more heavily on plant material and are less carnivorous than suids.. They locate food mainly by smell.. The tusks grow outwards or slightly downwards, never upward as in suids. Peccaries are also called Javelinas, a name derived from Spanish, meaning spear or Javelin, which refers to the spearlike tusks. Peccaries have a musk gland over the spine about 8 inches in front of the tail, which emits a very strong odor.Peccaries live in a variety of habitats from deserts to dense tropical forests.They are indigenous to the New World and range from southwestern United States to central Argentina. Gregarious animals, some may form bands of several dozen. The young are highly precocial and are able to run a few hours after birth. Peccaries are hunted for their meat and hide.

Catagonus wagneri   Chacoan peccary (V)
O   Bolivia, Paraguay,Argentina
     Z   W   P.W
                                                        P.Z

Tayassu albirostris   White lipped peccary.
O   Mexico to Argentina
     Z   W   P.W
                                                        P.Z

Tayassu tajacu   Collared peccary
O   Arizona and Texas to Argentina
     Z   W   P.W

                                                        P.Z

FAMILY: HIPPOPOTAMIDAE, Hippopotamuses

Hippopotami are large animals, with big heads and short limbs, that live in Africa in or near water.. They have evolved many adaptations which make them well suited for an amphibious life. The skin has specialized glands, which secrete a pink oily substance to protect the sparsely haired body. The eyes, ears, and nostrils, especially of the more aquatic river hippo, are located on top of the head, enabling the animal to see, breathe, and hear ,without raising the body out of the water. The incisors and canines are tusk-like and ever-growing. The lower incisors project outward; the lower canines are large and grow upward. The tusks are used for making artifacts. The huge river hippos (weighing up to 7000lbs.), are gregarious; the groups spend much of the day in the water. They are good swimmers and divers, and are able to walk on the bottom of the river or a lake. At night they move onto land to feed on vegetation. The pygmy hippo is less aquatic and less gregarious, occurring singly or in pairs in densely forested areas.

Choeropsis liberiensis		Pygmy hippopotamus (V)	
O Siera Leone to Nigeria			
Z	W		P.W
			P.Z
Hippopotamus amphibius		Hippopotamus	
O Africa			
Z	W		P.W
			P.Z

The following families belong to the sub-order RUMINANTIA. The members are herbivorous and well adapted to obtain maximum nutritional value from their food. Ruminants chew their cud- i.e., they chew, swallow, regurgitate, re-chew and re-swallow. In the first two chambers of the stomach, food is softened by muscular action; cellulose broken down by microorganisms. In the last chamber (third or fourth depending on the family), gastric digestion takes place. The teeth are well adapted to the grazing diet- - the molars have crescent shaped cusps (selenodont), which aid in grinding vegetation. The upper incisors are either absent or reduced; the upper canines are usually absent. The limbs are adapted for cursorial (running) locomotion. The bones of the lower limbs and especially those of the hands and feet are lengthened, with fusion taking place in the carpals and tarsals, (wrist and ankle bones) .There is no clavicle, which aids in lengthening the stride. With the exception of family Tragulidae, ruminants have a cannon bone on all four feet- - a single bone formed by the lengthening and fusion of the 3 rd. and 4 th. metapodials, (foot bones). With the exception of the camel, all ruminants stand on the tip of the hooved toes (unguligrade),and adaptation for speed.

FAMILY: CAMELIDAE, Camels, Guanaco, Llama, Alpaca, Vicuna

Camelids are rather large animals, with long necks and long limbs, well adapted to living in arid areas. Unlike other artiodactyls, which walk on the tips of their hooved toes (unguligrade), camelids walk on the toes themselves (digitigrade), which have nails rather than hooves. The two toes are separate, and each is supported by a broad pad, which increases the surface area of the foot, giving better support in soft, sandy soil.The feet are broad in camels and slender in the other family members.. These specializations of the feet evolved secondarily, (primitive camels were nearly unguligrade), possibly as adaptations to increasingly arid conditions. With the exception of the foot, camelids are the most primitive of the living ruminants, having a 3 chambered stomach and less specialized dentition.Camels can go for long periods without water due to many physiological adaptations .The kidneys, for example,are able to concentra urine, reducing water loss, and water is reabsorbed from fecal material in the intestional tract. The camel also can tolerate tremendous losses of weight through water loss but can recover the weight very quickly, by drinking large quantities of water without ill effects.Camelids live in widely separated geographical areas. Domesticated dromedary (one humped), and bactrian (two humped), camels, range from northern Africa through central Asia. Wild bactrian camel herds persist in the Gobi desert of Asia. Wild Guanacos and Vicunas live in herds throughout the Andes of the New World.Alpacas and Llamas inhabit the same areas but only in the domesticated state.

Camelus bactrianus		Camel (V)	
O Soviet Central Asia to Mongolia			
Z	W		P.W
			P.Z

Camelus dromedius   Dromedary
O   Probably once found throughout the Arabian region
Z   P

Lama glama   Llama
O   Found in domestication from Peru to Argentina
Z   P

Lama guanicoe   Guanaco
O   S. America,(Andes)
Z   W   P.W

                                                P.Z

Lama pacos   Alpaca
O   Found in domestication in Peru and Bolivia
Z   P

Lama vicugna   Vicuna (V)
O   Peru to Chile
Z   W   P.W

                                              P.Z

FAMILY: TRAGULIDAE, Mouse deer or Chevrotains

Tragulids are very small animals, weighing from 4.5 to 10 lbs., which live in tropical forests of Central Africa and parts of Southeast Asia. They probably resemble the ancestors of the more advanced artiodactyls (bovids, cervids,etc..) They have no horns or antlers, but have large upper canines which are used as defensive weapons.The limbs are long and slender,with the lateral digits (2 and 5) retained and fully developed in hind limbs only.The metacarpals (bones of fore feet), remain separate in the African form, and are partly fused in Asian forms. The ruminating stomach is 3 chambered- - the diet consists of grass, leaves, and some fruit.Little is known of these nocturnal, secretive creatures. They live in thick underbrush and dart into dense vegetation when threatened.

Hyemoschus aquaticus   Water chevrotain
O   Sierra Leone to Uganda
Z   W   P.W
                                            P.Z

Tragulus javanicus   Small Malayan chevrotain
O   S.E. Asia
Z   W   P.W
                                            P.Z

Tragulus meminna   Indian chevrotain
O   India, Sri Lanka
Z   W   P.W
                                            P.Z

Tragulus napu   Large Malayan chevrotain
O   S.E.Asia
Z   W   P.W
                                            P.Z

Cervids are found from the arctic, to the tropics; throughout most of the New World, Eurasia, and Northwest Africa. The most obvious distinguishing feature of this family, is the sexual ornament weapons worn on the heads of the males- - the antlers. Antlers are present in all but two genera of cervids, and are often the principal external means of identifying a species. In the genus Rangifer (reindeer and caribou), both male and female bear antlers.A distinct growth cycle is seen in antlers, which are shed and regrown annually. In temperate zones, the growth begins in early summer. From a short base on the frontal bones of the skull, called the pedicel, bone growth begins; - soft, tender bone,well supplied with blood and covered with a thin skin and soft, fine hairs (the velvet). Late in summer when the antlers' maximum growth is attained, the blood supply recedes, the hair and skin die, and are rubbed off. During the ensuing mating season, the bulls may fight in ritual combat for the females, to form a harem. Antlers are used in charging each other and may become interlocked, resulting in the death by starvation of both bulls. Following the mating season the antlers are shed.There are other features which are found in cervids. Among them are preorbital (in front of the eye), glands on the face, which occur in nearly all deer, and glands on the hind limbs. The feet of deer have 4 toes, but the lateral digits are often greatly reduced in size. An interesting correlation occurs between antlers and canine teeth:- Canines are usually absent or reduced in size in antlered species.. In two genera which have very short underdeveloped antlers, the canines are enlarged, and in the two genera that carry no antlers the canines are saber-like.Cervidae range in size from about 22lbs to 1750 lbs .Most are gregarious and typically are harem breeders. Some are solitary. Gestation varies greatly. These hervibores are browsers, eating the leaves, shoots, branches, and if necessary,bark of trees and shrubs. The stomach is four chambered and ruminating.

Moschus moschiferus			Musk deer (E*)		
O	Asia				
	Z	W		P.W	
				P.Z	
Elaphodus cephalophus			Tufted deer		
O	China, Burma				
	Z	W		P.W	
				P.W	
Muntiacus crinifrons			Black muntjac (I)		
O	E.China				
	Z	W		P.W	
				P.Z	
Muntiacus feae			Feas muntjac (I)		
O	Thailand				
	Z	W		P.W	
				P.Z	
Muntiacus muntjak			Muntjac		
O	India, Napal to China,S.E.Asia				
	Z	W		P.W	
				P.Z	
Muntiacus reevesi			Reeves muntjac		
O	S.E.China,Taiwan				
	Z	W		P.W	
				P.Z	
Muntiacus rooseveltorum			Roosevelt's muntjac		
O	Laos				
	Z	W		P.W	
				P.Z	

Cervus albirostris	Thorold's deer (I)		
O Tibet, W.China			
Z	W		P.W
			P.Z
Cervus axis	Axis deer		
O India, Sri Lanka, Nepal			
Z	W		P.W
			P.Z
Cervus dama	Fallow deer		
O Europe, Asia minor			
Z	W		P.W
			P.Z
Cervus duvauceli	Barasingha (E)		
O India			
Z	W		P.W
			P.Z
Cervus elaphus	Red deer		
O Europe,Asia, N.Africa, N.America			
Z	W		P.W
			P.Z
Cervus eldi	Eld,s deer (E*)		
O Burma, Thailand, Viet Nam			
Z	W		P.W
			P.Z
Cervus mariannus	Philippine sambar		
O Philippines			
Z	W		P.W
			P.Z
Cervus mesopotamicus	Persian fallow deer		
O Iran			
Z	W		P.W
			P.Z
Cervus nippon	Sika deer (E)		
O S.Asia, Japan, Taiwan			
Z	W		P.W
			P.Z
Cervus porcinus	Hog deer (E*)		
O S.Asia			
Z	W		P.W
			P.Z
Cervus timorensis	Timor deer		
O Java,Celebes, Timor			
Z	W		P.W
			P.Z
Cervus unicolor	Sambar		
O India to China and S.E.Asia			
Z	W		P.W
			P.Z

Elaphurus davidianus	Pere David's deer	
O Zoos only (original habitat China)		
Z		P
Alces alces	Moose	
O N.Europe,N.Asia, N.America		
Z	W	P.W
		P.Z
Blastocerus campestris	Pampas deer	
O Brazil, Paraguay, Argentina, Uruguay		
Z	W	P.W
		P.Z
Blastocerus dichotomus	Marsh deer (V)	
O Brazil, N.Argentina		
Z	W	P.W
		P.Z
Capreolus capreolus	Roe deer	
O Europe, Asia		
Z	W	P.W
		P.Z
Hippocamelus antisensis	Peruvian guemal (V)	
O Andes, Ecuador to Argentina		
Z	W	P.W
		P.Z
Hippocamelus bisulcus	Chilean guemal (V)	
O Andes, Chile , Argentina		
Z	W	P.W
		P.Z
Hydropotes inermis	China water deer	
O China,Korea		
Z	W	P.W
		P.Z
Mazama americana	Red brocket deer	
O Mexico to Argentina		
Z	W	P.W
		P.Z
Mazama bricenii	Brocket deer	
O W.Venezuela		
Z	W	P.W
		P.Z
Mazama gouazoubira	Brown brocket deer	
O Mexico to N.South America		
Z	W	P.W
		P.Z
Mazama nana	Small brocket deer	
O South America		
Z	W	P.W
		P.Z.

Odocoileus hemoinus   Mule deer
O   W.and C. North America to N.Mexico
       Z   W   P.W

                                                                    P.Z
Odocoileus virginianus   Whitetail deer
O   North to South America
       Z   W   P.W

                                                                    P.Z
Pudu mephistophiles   Pudu (I)
O   Colombia,Peru, Ecuador
       Z   W   P.W
                                                                    P.Z
Pudu pudu   Pudu (E*)
O   Chile, Argentina
       Z   W   P.W
                                                                    P.Z
Rangifer tarandus   Relndeer or Caribou
O   N.Europe, N.Asia, N.America

FAMILY: GIRAFFIDAE,Giraffes and Okapi

Giraffes and okapis are large, long-necked, long legged animals. These characteristics are
extreme in the giraffe, the tallest of all mammals (to 18 feet) . The eyes and ears are large;
the eyes are set well to the side, extending the field of vision. Giraffids have 2-5 horns
which are short, composed entirely of bone, and covered with haired skin. They occur in
both sexes of giraffe, and on male okapi, and are never shed. The fur of the giraffe has
brownish blotches of various sizes and shapes, on a buff ground color. Okapis are
maroon or purplish, with white and black stripes on the buttocks and upper portion of the
limbs. The low crowned teeth, adapted to a browsing diet, are covered with enamel that is
"wrinkled" (rugose),rather than smooth as in all other browsers. The stomach is 4
chambered and ruminating.Giraffids inhabit much of Africa south of the Sahara. Giraffes
live on the savannahs, feeding on high branches of the acacia and other trees, using their
long and flexible tongue for breaking off the leaves. They live in loosely structured herds
of different compositions. They run when in danger, but can defend themselves by
kicking. They are vulnerable to lions when spreading the forelegs and lowering the head
to drink. Okapis live in dense tropical forests and feed on leaves and fruit. They are solitary
or occur in pairs or small family groups- - never in herds. Okapi were unknown until 1900.

Giraffa camelopardalis   Giraffe
O   Africa
       Z   W   P.W

                                                                    P.Z
Okapia johnstoni   Okapi
O   Zaire
       Z   W   P.W

                                                                    P.Z

FAMILY: ANTILOCAPRIDAE, Pronghorn

The distinguishing feature of the pronghorn is its horn, which differentiates it from antelopes of the family Bovidae. The core of the horn is permanent, and composed of true bone which grows from the skull. It is covered with a hollow black outer sheath composed of true horn, like that of bovids, but unlike bovid horn, it is branched and shed annually. Both sexes have horns, but those of females are small and inconspicuous.The pronghorn is well adapted for flight, having long slender limbs and hooves cushioned with cartilaginous padding, which is larger in the forefeet than in hind feet; probably to cushion the impact of landing on the forefeet while running. The eyes are set far back and to the side of the skull providing a wide field of vision.Pronghorns live in open prairies and deserts of western North America, from Canada to Mexico. The fossil record of pronghorns is entirely North American. They browse on low shrubs, and forbs,(any herb that is not a grass),and also eat some grass. The teeth are high crowned, uncommon for a browser, but an advantage, due to the large amount of soil particles in the low vegetation. The stomach is 4 chambered and ruminating. Gregarious animals, pronghorn herds number from 50 to 100 individuals during winter. During spring fawning season, they break down into small bands. In fall, males fight for harems of up to 15 does. Subsequently, she normally gives birth to twins.

Antilocapra americana   Pronghorn (E*)
O   South West United States to Mexico
      Z   W   P.W

                                                            P.Z

FAMILY: BOVIDAE, Antelope, Cattle, Buffalo, Goats, Sheep

This family is an extremely diverse and important group. It represents the major food source for man and other large carnivores. Nearly world wide in distribution, only south America, Australia and Antarctica, lack native wild inhabitants. While "typically" grassland animals, they occupy many types of terrain in many climates. The oryx and addax live in very arid areas; water buffalo inhabit wet areas; musk-oxen survive in extreme cold; mountain goats and many sheep are adapted to live in high mountains; and forest duikers are found in dense jungle. Among the domesticated bovids, are true cattle, goats, and sheep.Members of the family bovidae are characterized by the presence of horns. Unlike antlers, horns grow throughout the life of an animal, and are never shed. They are made up of a bony core and an ever growing keratinized (horny) sheath. They may be simple spikes, or long and spectacularly curved or spiraled, but they are never branched.In most species, both males and females carry two horns, although there are species with 4 horns and some in which females have none.Bovids are very well adapted to a fugitive life. The limbs are long and slender for running. The 4 chambered ruminating stomach allows the animal to eat quickly, and retire to a safe place, to rechew the food. The large eyes are set to the side of the head to extend the field of vision. Young are born in a precocial state and are capable of standing and running soon after birth.Most bovids are grazers. The teeth are high crowned to deal with the soil particles in the grasses. In the grasslands of Africa where herds of many species overlap, specialization in feeding has occurred. To the extent, that some species feed only on the older, taller grasses; some on the lower portions of those same grasses and others only on low growing young shoots. Some species require little or no access to drinking water.

308

Bison bison   American bison(E*)
O   N.America
    Z   W   P.W

                                                        P.Z

Bison bonasus   Wisent (E*)
O   E.Europe
    Z   W   P.W
                                                        P.Z

Bos gaurus   Gaur (V)
O   India to Malay Peninsula
    Z   W   P.W
                                                        P.Z

Bos javanicus   Banteng (V)
O   S.Asia,Java,Borneo
    Z   W   P.W
                                                        P.Z

Bos mutus   Wild yak
O   Northern Kansu
    Z   W   P.W
                                                        P.Z

Bos grunniens   Domestic yak (E)
O   Tibet
    Z   W   P.W
                                                        P.Z

Bos primigenius   Wild ox
O   Europe, Asia, N.Africa
    Z   W   P.W
                                                        P.Z

Bos sauveli   Kouprey (E*)
O   Indochina
    Z   W   P.W
                                                        P.Z

Bos taurus   **Domestic cattle**
O   **world wide**

Boselaphus tragocamelus   Nilgai
O   Pakistan, India
    Z   W   P.W
                                                        P.Z

Bubalus arnee   Asiatic buffalo (V)
O   Asia
    Z   W   P.W
                                                        P.Z

Bubalus depressicornis   Anoa (E)
O   Celebes
    Z   W   P.W
                                                        P.Z

Bubalus mindorensis			Tamarau (E)
O Philippines			
Z	W		P.W
			P.Z
Bubalus quarlesi			Mountain anoa (E)
O Highlands of Celebes			
Z	W		P.W
			P.Z
Syncerus caffer			African buffalo
O Africa			
Z	W		P.W
			P.Z
Taurotragus derbianus			Giant eland (E)
O Senegal to Sudan			
Z	W		P.W
			P.Z
Taurotragus oryx			Common eland
O E.and S.Africa			
Z	W		P.W
			P.Z
Tetracerus quadricornis			Four horned antelope
O India, Nepal			
Z	W		P.W
			P.Z
Tragelaphus angasi			Nyala
O Africa			
Z	W		P.W
			P.Z
Tragelaphus buxtoni			Mountain nyala
O Ethiopia			
Z	W		P.W
			P.Z
Tragelaphus euryceros			Bongo
O Sierra Leone to Kenya			
Z	W		P.W
			P.Z
Tragelaphus imberbis			Lesser kudu
O E.Africa			
Z	W		P.W
			P.Z
Tragelaphus scriptus			Bushbuck
O Africa south of the Sahara			
Z	W		P.W
			P.Z

Tragelaphus spekei   Sitatunga
O   Gambia to Sudan, Botswana
     Z   W   P.W

                                              P.Z

Tragelaphus strepsiceros   Greater kudu
O   C.andS.Africa
     Z   W   P.W
                                              P.Z.

Cephalophus adersi   Zanzibar duiker
O   Kenya, Zanzibar
     Z   W   P.W
                                              P.Z

Cephalophus callipygus   Peter's duiker
O   Cameroon,Gabon, Congo
     Z   W   P.W
                                              P.Z

Cephalophus dorsalis   Bay duiker
O   W.Africa to Zaire
     Z   W   P.W
                                              P.Z

Cephalophus jentinki   Jentink's duiker (E)
O   Liberia
     Z   W   P.W
                                              P.Z

Cephalophus leucogaster   Gaboon duiker
O   Cameroon to Zaire
     Z   W   P.W
                                              P.Z

Cephalophus maxwelli   Maxwell's duiker
O   Senegal to Nigeria
     Z   W   P.W
                                              P.Z

Cephalophus monticola   Blue duiker
O   Nigeria to Kenya to South Africa
     Z   W   P.W
                                              P.Z

Cephalophus natalensis   Red duiker
O   Zaire, Somalia to South Africa
     Z   W   P.W
                                              P.Z

Cephalophus niger   Black duiker
O   Guinea to Nigeria
     Z   W   P.W
                                              P.Z

Cephalophus nigrifrons   Black fronted duiker
O   Cameroon to Kenya, Angola
     Z   W   P.W
                                              P.Z

Cephalophus ogilbyi   Ogilby's duiker
O   Siera Leone to Cameroon
         Z   W   P.W
                                                         P.Z

Cephalophus rufilatus   Red flanked duiker
O   Senegal to Sudan
         Z   W   P.W
                                                         P.Z

Cephalophus spadix   Abbott's duiker
O   Tanzania
         Z   W   P.W
                                                         P.Z

Cephalophus sylvicultor   Yellow backed duiker
O   Gambia to Kenya,Angola,Zambia
         Z   W   P.W
                                                         P.Z

Cephalophus zebra   Zebra banded duiker
O   Sierra Leone to Ivory Coast
         Z   W   P.W
                                                         P.Z

Sylvicapra grimmia   Common duiker
O   Africa
         Z   W   P.W
                                                         P.Z

Addax nasomaculatus   Addax (V)
O   Sahara
         Z   W   P.W
                                                         P.Z

Alcelaphus buselaphus   Hartebeest (E)
O   Africa
         Z   W   P.W
                                                         P.Z

Alcelaphus lichtensteini   Lichtenstein's hartebeest
O   E.to S.Africa
         Z   W   P.W
                                                         P.Z

Connochaetes gnou   Black wildebeest
O   South Africa
         Z   W   P.W
                                                         P.Z

Connochaetes taurinus   Blue wildebeest
O   Kenya to South Africa
         Z   W   P.W
                                                         P.Z

Damaliscus dorcas   Bontebok (E*)
O   South Africa
         Z   W   P.W
                                                         P.Z

Damaliscus hunteri   Hunter's hartebeest (R)
O   Kenya, Somalia
         Z   W   P.W
                                                         P.Z

Damaliscus lunatus   Topi
O   Senegal to Ethiopia and South Africa
     Z   W   P.W
                                                        P.Z

Hippotragus equinus   Roan antelope
O   Senegal to Ethiopia to South Africa
     Z   W   P.W
                                                        P.Z

Hippotragus niger   Sable antelope (E)
O   Kenya to South Africa
     Z   W   P.W
                                                        P.Z

Kobus ellipsiprymnus   Waterbuck
O   Savannah zones of Africa
     Z   W   P.W
                                                        P.Z

Kobus kob   Kob
O   Senegal to Kenya
     Z   W   P.W
                                                        P.Z

Kobus leche   Lechwe (V)
O   Zaire to South Africa
     Z   W   P.W
                                                        P.Z

Kobus megaceros   Nile lechwe
O   Swamp of Sudan and Ethiopia
     Z   W   P.W
                                                        P.Z

Kobus vardoni   Puku
O   Zaire,Tanzania,Angola,Zambia
     Z   W   P.W
                                                        P.Z

Oryx dammah   Scimitar oryx (V)
O   Morocco and Senegal to Egypt and Sudan
     Z   W   P.W
                                                        P.Z

Oryx gazella   Oryx
O   E. to S.Africa
     Z   W   P.W
                                                        P.Z

Oryx leucoryx   Arabian oryx (E)
O   Arabian peninsula
     Z   W   P.W
                                                        P.Z

Pelea capreolus   Grey rhebok
O   South Africa
     Z   W   P.W
                                                        P.Z

Redunca arundinum   Reedbuck
O   Gabon,Tanzania to South Africa
     Z   W   P.W
                                                        P.Z

Redunca fulvorufula   Mountain reedbuck
O   Cameroon,E.and S.Africa
     Z   W   P.W
     P.Z

Redunca redunca   Bohor reedbuck
O   Senegal to Ethiopia ,Tanzania
     Z   W   P.W
     P.Z

Aepyceros melampus   Impala
O   E.and S.Africa
     Z   W   P.W
     P.Z

Ammodorcas clarkei   Dibatag (V)
O   Somalia, Ethiopia
     Z   W   P.W
     P.Z

Antidorcas marsupialis   Springbok
O   S.Africa
     Z   W   P.W
     P.Z

Antilope cervicapra   Blackbuck
O   Pakistan,India
     Z   W   P.W
     P.Z

Dorcatragus megalotis   Beira antelope
O   Somalia, Ethiopia
     Z   W   P.W
     P.Z

Gazella cuvieri   Edmi gazelle (E)
O   Atlas Mts. of North Africa
     Z   W   P.W
     P.Z

Gazella dama   Dama gazelle (E)
O   Morocco, Senegal to Sudan
     Z   W   P.W
     P.Z

Gazella dorcas   Dorcas gazelle (E)
O   Africa to India
     Z   W   P.W
     P.Z

Gazella gazella   Mountain gazelle (E*)
O   Arabian Peninsula
     Z   W   P.W
     P.Z

Gazella granti   Grant's gazelle
O   E. Africa
     Z   W   P.W
     P.Z

Gazella leptoceros   Slender horned gazelle (E)
O   Algeria to Egypt
     Z   W   P.W
     P.Z

314

Gazella rufifrons   Red fronted gazelle
O   Senegal to Ethiopia
    Z   W   P.W
                                            P.Z

Gazella soemmeringi   Soemmering's gazelle
O   Ethiopia,Sudan,Somalia
    Z   W   P.W
                                            P.Z

Gazella spekei   Speke's gazelle (I)
O   Ethiopia, Somalia
    Z   W   P.W
                                            P.Z

Gazella subgutturosa   Goitered gazelle (E)
O   Palestine,Arabian Peninsula, China
    Z   W   P.W
                                            P.Z

Gazella thomsoni   Thomson's gazelle
O   Sudan, Kenya,Tanzania
    Z   W   P.W
                                            P.Z

Litocranius walleri   Gerenuk
O   Somalia,Kenya,Ethiopia
    Z   W   P.W
                                            P.Z

Madoqua guentheri   Guenther's dik-dik
O   C.and E.Africa
    Z   W   P.W
                                            P.Z

Madoqua kirki   Kirk's dik-dik
O   E.Africa to Namibia
    Z   W   P.W
                                            P.Z

Madoqua phillipsi   Phillip's dik-dik
O   Ethiopia, Somalia
    Z   W   P.W
                                            P.Z

Madoqua saltiana   Salt's dik-dik
O   Ethiopia,Sudan, Somalia
    Z   W   P.W
                                            P.Z

Madoqua swaynei   Swaynes dik-dik
O   Ethiopia,Somali
    Z   W   P.W
                                            P.Z

Neotragus batesi   Bate's dwarf antelope
O   Nigeria to Uganda
    Z   W   P.W
                                            P.Z

Neotragus moschatus   Suni (E)
O   Kenya to South Africa
    Z   W   P.W
                                            P.Z

Neotragus pygmaeus		Royal antelope	
O Sierra Leone to Ghana			
Z	W		P.W
			P.Z
Oreotragus oreotragus		Klipspringer	
O C.E.and S.Africa			
Z	W		P.W
			P.Z
Ourebia ourebi		Oribi	
O Africa south of the Sahara			
Z	W		P.W
			P.Z
Procapra gutturosa		Mongolia gazelle	
O Siberia,Mongolia, China			
Z	W		P.W
			P.Z
Procapra picticaudata		Tibetan gazelle	
O Mongolia, Tibet,China			
Z	W		P.W
			P.Z
Raphicerus campestris		Steenbok	
O Kenya to South Africa			
Z	W		P.W
			P.Z
Raphicerus melanotis		Grysbok	
O Coastal South Africa			
Z	W		P.W
			P.Z
Raphicerus sharpei		Sharpe's grysbok	
O Tanzania to South Africa			
Z	W		P.W
			P.Z
Ammotragus lervia		Aoudad or Barbary sheep	
O N. Africa			
Z	W		P.W
			P.Z
Budorcas taxicolor		Takin	
O C.China			
Z	W		P.W
			P.Z
Capra aegagrus		Wild goat	
O Asia			
Z	W		P.W
			P.Z
Capra falconeri		Markhor (E*)	
O Afghanistan, Pakistan			
Z	W		P.W
			P.Z

Capra hircus   **Domestic goat**
O   Worldwide in association with man

Capra ibex   Ibex
O   Europe,Asia,N.Africa
    Z   W   P.W
                                    P.Z

Capra pyrenaica   Spanish ibex (E)
O   Spain, Portugal
    Z   W   P.W
                                    P.Z

Capra walie   Abyssinian ibex (E)
O   Ethiopia
    Z   W   P.W
                                    P.Z

Capricornis cripus   Japanese serow
O   Japan,Taiwan
    Z   W   P.W
                                    P.Z

Capricornis sumatraensis   Serow (E*)
O   S.E.Asia
    Z   W   P.W
                                    P.Z

Hemitragus hylocrius   Nilgiri tahr (V)
O   India
    Z   W   P.W
                                    P.Z

Hemitragus jayakari   Arabian tahr (E)
O   Oman
    Z   W   P.W
                                    P.Z

Hemitragus jemlahicus   Himalayan tahr
O   Himalayan region
    Z   W   P.W
                                    P.Z

Nemorhaedus goral   Goral (E*)
O   Siberia to Tibet,Thailand
    Z   W   P.W
                                    P.Z

Oreamnos americanus   Rocky mountain goat
O   Alaska,W.Canada,N.W.United States
    Z   W   P.W

                                    P.Z

Ovibos moschatus   Musk ox
O   N.Alaska and Canada
    Z   W   P.W

                                    P.Z

Ovis ammon   Argali
O   Himalayan region
    Z   W   P.W
                                    P.Z

Ovis aries   **Domestic sheep**
O   Worldwide in association with man

Ovis canadensis		Bighorn sheep		
O W.Canada to Mexico				
Z	W		P.W	
			P.Z	
Ovis dalli		Dall's sheep		
O Alaska,N.W.Canada				
Z	W		P.W	
			P.Z	
Ovis musimon		Mouflon		
O Corsica,Sardinia				
Z	W		P.W	
			P.Z	
Ovis nivicola		Snow sheep		
O N.E.Siberia				
Z	W		P.W	
			P.Z	
Ovis orientalis		Asiatic mouflon		
O Asia Minor, Iran				
Z	W		P.W	
			P.Z	
Pantholops hodgsoni		Tibetan antelope		
O Tibet,China				
Z	W		P.W	
			P.Z	
Pseudois nayaur		Bharal		
O Mongolia to Tibet				
Z	W		P.W	
			P.Z	
Rupicapra rupicapra		Chamois (E*)		
O S.Europe to the Caucasus				
Z	W		P.W	
			P.Z	
Saiga tatarica		Saiga antelope (E*)		
O Ukraine to Mongolia				
Z	W		P.W	
			P.Z	

TOTAL ARTIODACTYLA : O	Z	W	P.W	P.Z

UNITED STATES CHECK LIST

ALABAMA				
O	Z	W	P.W	P.Z

ALASKA				
O	Z	W	P.W	P.Z

ARIZONA				
O	Z	W	P.W	P.Z

ARKANSAS				
O	Z	W	P.W	P.Z

CALIFORNIA				
O	Z	W	P.W	P.Z

COLORADO				
O	Z	W	P.W	P.Z

CONNECTICUT				
O	Z	W	P.W	P.Z

DELAWARE				
O	Z	W	P.W	P.Z

UNITED STATES CHECK LIST

		DISTRICT OF COLUMBIA		
O	Z	W	P.W	P.Z

		FLORIDA		
O	Z	W	P.W	P.Z

		GEORGIA		
O	Z	W	P.W	P.Z

		HAWAII		
O	Z	W	P.W	P.Z

		IDAHO		
O	Z	W	P.W	P.Z

		ILLINOIS		
O	Z	W	P.W	P.Z

		INDIANA		
O	Z	W	P.W	P.Z

		IOWA		
O	Z	W	P.W	P.Z

UNITED STATES CHECK LIST

KANSAS

| O | Z | W | P.W | P.Z |

KENTUCKY

| O | Z | W | P.W | P.Z |

LOUISIANA

| O | Z | W | P.W | P.Z |

MAINE

| O | Z | W | P.W | P.Z |

MARYLAND

| O | Z | W | P.W | P.Z |

MASSACHUSETTS

| O | Z | W | P.W | P.Z |

MICHIGAN

| O | Z | W | P.W | P.W. |

MINNESOTA

| O | Z | W | P.W | P.Z |

UNITED STATES CHECK LIST

		MISSISSIPPI		
O	Z	W	P.W	P.Z

		MISSOURI		
O	Z	W	P.W	P.Z

		MONTANA		
O	Z	W	P.W	P.Z

		NEBRASKA		
O	Z	W	P.W	P.Z

		NEVADA		
O	Z	W	P.W	P.Z

		NEW HAMPSHIRE		
O	Z	W	P.W	P.Z

		NEW JERSEY		
O	Z	W	P.W	P.Z

		NEW MEXICO		
O	Z	W	P.W	P.Z

UNITED STATES CHECK LIST

NEW YORK

O　　　　Z　　　　W　　　　P.W　　　　P.Z

NORTH CAROLINA

O　　　　Z　　　　W　　　　P.W　　　　P.Z

NORTH DAKOTA

O　　　　Z　　　　W　　　　P.W　　　　P.Z

OHIO

O　　　　Z　　　　W　　　　P.W　　　　P.Z

OKLAHOMA

O　　　　Z　　　　W　　　　P.W　　　　P.Z

OREGON

O　　　　Z　　　　W　　　　P.W　　　　P.Z

PENNSYLVANIA

O　　　　Z　　　　W　　　　P.W　　　　P.Z

RHODE ISLAND

O　　　　Z　　　　W　　　　P.W　　　　P.Z

UNITED STATES CHECK LIST

		SOUTH CAROLINA		
O	Z	W	P.W	P.Z

		SOUTH DAKOTA		
O	Z	W	P.W	P.Z

		TENNESSEE		
O	Z	W	P.W	P.Z

		TEXAS		
O	Z	W	P.W	P.Z

		UTAH		
O	Z	W	P.W	P.Z

		VERMONT		
O	Z	W	P.W	P.Z

		VIRGINIA		
O	Z	W	P.W	P.Z

		WASHINGTON		
O	Z	W	P.W	P.Z

UNITED STATES CHECK LIST

WEST   VIRGINIA

O   Z   W   P.W   P.Z

WISCONSIN

O   Z   W   P.W   P.Z

WYOMING

O   Z   W   P.W   P.Z

TOTAL : O	Z	W	P.W	P.Z

CANADA CHECK LIST

ALBERTA

O   Z   W   P.W   P.Z

BRITISH COLUMBIA

O   Z   W   P.W   P.Z

MANITOBA

O   Z   W   P.W   P.Z

NEW BRUNSWICK

O   Z   W   P.W   P.Z

NEW FOUNDLAND AND LABRADOR

O   Z   W   P.W   P.Z

NORTHWEST TERRITORIES

O   Z   W   P.W   P.Z

NOVA SCOTIA

O   Z   W   P.W   P.Z

ONTARIO

O   Z   W   P.W   P.Z

CANADA CHECK LIST

PRINCE EDWARD ISLAND

O	Z	W	P.W	P.W.

QUEBEC

O	Z	W	P.W	P.Z

SASKATCHEWAN

O	Z	W	P.W	P.Z

YUKON

O	Z	W	P.W	P.Z

TOTAL : O	Z	W	P.W	P.Z

COUNTRIES CHECK LIST

O   Z   W   P.W   P.Z

AFARS & ISSAS,

O   Z   W   P.W   P.Z

AFGHANISTAN, AFG.

O   Z   W   P.W   P.Z

ALBANIA, ALB.

O   Z   W   P.W   P.Z

ALGERIA

O   Z   W   P.W   P.Z

AMERICAN SAMOA,

O   Z   W   P.W   P.Z

ANDORRA, AND.

O   Z   W   P.W   P.Z

ANGOLA, ANG.

O   Z   W   P.W   P.Z

ARGENTINA, ARG.

O   Z   W   P.W   P.Z

AUSTRALIA, AUSTL.

O   Z   W   P.W   P.Z

AUSTRIA, AUS.

O   Z   W   P.W   P.Z

BAHAMAS, BA.

O   Z   W   P.W   P.Z

BAHRAIN,

O   Z   W   P.W   P.Z

332

COUNTRIES CHECK LIST

BANGLADESH, BNGL.

O   Z   W   P.W   P.Z

BELGIUM, BEL.

O   Z   W   P.W   P.Z

BELIZE,

O   Z   W   P.W   P.Z

BENIN,

O   Z   W   P.W   P.Z

BERMUDA,

O   Z   W   P.W   P.Z

BHUTAN, BHU

O   Z   W   P.W   P.Z

BOLIVIA, BOL.

O   Z   W   P.W   P.Z

BOTSWANA, BOTS.

O   Z   W   P.W   P.Z

BRAZIL, BRAZ.

O   Z   W   P.W   P.Z

BRUNEI, BRU.

O   Z   W   P.W   P.Z

COUNTRIES CHECK LIST

BULGARIA, BUL.

O	Z	W	P.W	P.Z

BURMA, BUR.

O	Z	W	P.W	P.Z

BURUNDI,

O	Z	W	P.W	P.Z

CAMBODIA, CAMB.

O	Z	W	P.W	P.Z

CAMEROON, CAM.

O	Z	W	P.W	P.Z

CANADA, CAN.

O	Z	W	P.W	P.Z

CAPE VERDE, C.V.

O	Z	W	P.W	P.Z

CENTRAL AFRICAN REPUBLIC, CEN. AFR. REP.

O	Z	W	P.W	P.Z

CHAD,

O	Z	W	P.W	P.Z

CHILE,

O	Z	W	P.W	P.Z

COUNTRIES CHECK LIST

CHINA,

O   Z   W   P.W   P.Z

COLOMBIA,   COL.

O   Z   W   P.W   P.Z

COMORO IS.,

O   Z   W   P.W   P.Z

CONGO,CON.

O   Z   W   P.W   P.Z

COSTA   RICA,   C.R.

O   Z   W   P.W   P.Z

CUBA,

O   Z   W   P.W   P.Z

CYPRUS,   CYP.

O   Z   W   P.W   P.Z

CZECHOSLOVAKIA,   CZECH.

O   Z   W   P.W   P.Z

DENMARK,   DEN.

O   Z   W   P.W   P.Z

DJIBOUTI

O   Z   W   P.W   P.Z

DOMINICA

O   Z   W   P.W   P.Z

DOMINICAN   REPUBLIC,   DOM.   REP.

O   Z   W   P.W   P.Z

COUNTRIES CHECK LIST

<u>ECUADOR, EC</u>.

O   Z   W   P.W   P.Z

<u>EGYPT, EG</u>.

O   Z   W   P.W   P.Z

<u>EL SALVADOR, SAL</u>.

O   Z   W   P.W   P.Z

<u>ENGLAND, ENG</u>.

O   Z   W   P.W   P.Z

<u>EQUATORIAL GUINEA, EQUAT. GUI</u>.

O   Z   W   P.W   P.Z

<u>ETHIOPIA, ETH</u>.

O   Z   W   P.W   P.Z

<u>FIJI</u>,

O   Z   W   P.W   P.Z

<u>FINLAND, FIN</u>.

O   Z   W   P.W   P.Z

<u>FRANCE, FR</u>.

O   Z   W   P.W   P.Z

COUNTRIES CHECK LIST

FRENCH   GUIANA, FR. GU.

| O | Z | W | P.W | P.Z |

FRENCH   POLYNESIA,

| O | Z | W | P.W | P.Z |

FRENCH   WEST   INDIES,

| O | Z | W | P.W | P.Z |

GABON,

| O | Z | W | P.W | P.Z |

GAMBIA,   GAM.

| O | Z | W | P.W | P.Z |

GERMAN   DEMOCRATIC   REPUBLIC,   G.D.R.

| O | Z | W | P.W | P.Z |

GERMANY,   FEDERAL REPUBLIC   OF.

| O | Z | W | P.W | P.Z |

GHANA,

| O | Z | W | P.W | P.Z |

GREECE,   GRC.

| O | Z | W | P.W | P.Z |

COUNTRIES CHECK LIST

<u>GRENADA,</u>

O	Z	W	P.W	P.Z

<u>GUATEMALA, GUAT</u>.

O	Z	W	P.W	P.Z

<u>GUINEA,</u>

O	Z	W	P.W	P.Z

<u>GUINEA--BISSAU,</u>

O	Z	W	P.W	P.Z

<u>GUYANA,</u>

O	Z	W	P.W	P.Z

<u>HAITI, HAI</u>.

O	Z	W	P.W	P.Z

<u>HONDURAS, HOND</u>.

O	Z	W	P.W	P.Z

<u>HUNGARY, HUNG</u>.

O	Z	W	P.W	P.Z

<u>ICELAND, ICE</u>.

O	Z	W	P.W	P.Z

<u>INDIA,</u>

O	Z	W	P.W	P.Z

COUNTRIES CHECK LIST

<u>INDONESIA,   INDON</u>.

O   Z   W   P.W   P.Z

<u>IRAN</u>,

O   Z   W   P.W   P.Z

<u>IRAQ</u>,

O   Z   W   P.W   P.Z

<u>IRELAND,   IRE</u>.

O   Z   W   P.W   P.Z

<u>ISRAEL,   ISR</u>.

O   Z   W   P.W   P.Z

<u>ITALY,   IT</u>.

O   Z   W   P.W   P.Z

<u>IVORY   COAST,   I.C</u>.

O   Z   W   P.W   P.Z.

<u>JAMAICA,   JAM.</u>

O   Z   W   P.W   P.Z.

<u>JAPAN,   JAP.</u>

O   Z   W   P.W   P.Z.

<u>JORDAN</u>,

O   Z   W   P.W   P.Z.

COUNTRIES CHECK LIST

			KASHMIR.	
O	Z	W	P.W	P.Z.

			KENYA, KEN	
O	Z	W	P.W	P.Z.

			KOREA, KOR.	
O	Z	W	P.W	P.Z

			KUWAIT, KUW.	
O	Z	W	P.W	P.Z

			LAOS.	
O	Z	W	P.W	P.Z

			LEBANON, LEB.	
O	Z	W	P.W	P.Z

			LESOTHO, LESO.	
O	Z	W	P.W	P.Z

			LIBERIA	
O	Z	W	P.W	P.Z

			LIBYA,	
O	Z	W	P.W	P.Z

			LIECHTENSTEIN, LIECH.	
O	Z	W	P.W	P.Z

			LUXEMBOURG, LUX.	
O	Z	W	P.W	P.Z

COUNTRIES CHECK LIST

MADAGASCAR, MAD.

O	Z	W	P.W	P.Z

MALAWI,

O	Z	W	P.W	P.Z

MALAYSIA, MALA.

O	Z	W	P.W	P.Z

MALDIVES,

O	Z	W	P.W	P.Z

MALI

O	Z	W	P.W	P.Z

MALTA,

O	Z	W	P.W	P.Z

MAURITANIA, MAUR.

O	Z	W	P.W	P.Z

MAURITIUS,

O	Z	W	P.W	P.Z

MEXICO, MEX.

O	Z	W	P.W	P.Z

MONACO,

O	Z	W	P.W	P.Z

MONGOLIA, MONG.

O	Z	W	P.W	P.Z

MOROCCO, MOR.

O	Z	W	P.W	P.Z

341

COUNTRIES CHECK LIST

<u>MOZAMBIQUE,   MOZ</u>.

O	Z	W	P.W	P.Z

<u>NAURU,</u>

O	Z	W	P.W	P.Z

<u>NEPAL,   NEP</u>.

O	Z	W	P.W	P.Z

<u>NETHERLANDS,   NETH.</u>

O	Z	W	P.W	P.Z

<u>NETHERLANDS   ANTILLES,</u>

O	Z	W	P.W	P.Z

<u>NEW   ZEALAND,   N.Z.</u>

O	Z	W	P.W	P.Z

<u>NICARAGUA,   NIC</u>.

O	Z	W	P.W	P.Z

<u>NIGER,</u>

O	Z	W	P.W	P.Z.

<u>NIGERIA,   NIG</u>.

O	Z	W	P.W	P.Z.

<u>NORWAY,   NOR</u>.

O	Z	W	P.W	P.Z.

<u>OMAN,   OM.</u>

O	Z	W	P.W	P.Z.

342

COUNTRIES CHECK LIST

PAKISTAN, PAK.

O   Z   W   P.W   P.Z.

PANAMA, PAN.

O   Z   W   P.W   P.Z.

PAPUA NEW GUINEA, PAP.N.GUI.

O   Z   W   P.W   P.Z.

PARAGUAY, PAR.

O   Z   W   P.W   P.Z.

PERU,

O   Z   W   P.W   P.Z.

PHILIPPINES, PHIL.

O   Z   W   P.W   P.Z.

POLAND, POL.

O   Z   W   P.W   P.Z.

PORTUGAL, PORT.

O   Z   W   P.W   P.Z.

PUERTO RICO,

O   Z   W   P.W   P.Z.

QATAR,

O   Z   W   P.W   P.Z.

ROMANIA, ROM.

O   Z   W   P.W   P.Z.

343

COUNTRIES CHECK LIST

RWANDA,

O   Z   W   P.W   P.Z.

SAN MARINO,

O   Z   W   P.W   P.Z.

SAUDI ARABIA, SAU.AR.

O   Z   W   P.W   P.Z.

SENEGAL, SEN.

O   Z   W   P.W   P.Z.

SEYCHELLES,

O   Z   W   P.W   P.Z.

SIERRA LEONE, S.L.

O   Z   W   P.W   P.Z.

SINGAPORE,

O   Z   W   P.W   P.Z.

SOMALIA, SOM.

O   Z   W   P.W   P.Z.

SOUTH AFRICA, S.AFR.

O   Z   W   P.W   P.Z.

SPAIN, SP.

O   Z   W   P.W   P.Z.

SRI LANKA,

O   Z   W   P.W   P.Z.

SUDAN, SUD.

O   Z   W   P.W   P.Z.

COUNTRIES CHECK LIST

SURINAM, SME.

O   Z   W   P.W   P.Z.

SWAZILAND, SWAZ.

O   Z   W   P.W   P.Z.

SWEDEN, SWE.

O   Z   W   P.W   P.Z.

SWITZERLAND, SWITZ.

O   Z   W   P.W   P.Z.

SYRIA, SYR.

O   Z   W   P.W   P.Z.

TAIWAN,

O   Z   W   P.W   P.Z.

TANZANIA, TAN.

O   Z   W   P.W   P.Z.

THAILAND, THAI.

O   Z   W   P.W   P.Z.

TOGO,

O   Z   W   P.W   P.Z.

TONGA,

O   Z   W   P.W   P.Z.

TRINIDAD & TOBAGO, TRIN.

O   Z   W   P.W   P.Z.

COUNTRIES CHECK LIST

O   Z   __TUNISIA,   TUN__.
     W   P.W   P.Z.

O   Z   __TURKEY,   TUR__.
     W   P.W   P.Z.

O   Z   __UGANDA,   UG__.
     W   P.W   P.Z.

O   Z   __U.S.S.R.,(SOVIET   UNION)__
     W   P.W   P.Z.

O   Z   __UNITED ARAB EMIRATES,   U.A.E__.
     W   P.W   P.Z.

O   Z   __UNITED STATES OF AMERICA, U.S.A__.
     W   P.W   P.Z.

O   Z   __UPPER   VOLTA,__
     W   P.W   P.Z.

O   Z   __URUGUAY,   UR__.
     W   P.W   P.Z.

O   Z   __VENEZUELA,   VEN__.
     W   P.W   P.Z.

COUNTRIES CHECK LIST

<u>VIET NAM, VIET</u>.

O   Z   W   P.W   P.Z.

<u>WESTERN SAMOA, W.SAM.</u>

O   Z   W   P.W   P.Z.

<u>YEMEN,</u>

O   Z   W   P.W   P.Z.

<u>YUGOSLAVIA, YUGO</u>.

O   Z   W   P.W   P.Z.

<u>ZAIRE,</u>

O   Z   W   P.W   P.Z.

<u>ZAMBIA,</u>

O   Z   W   P.W   P.Z.

<u>ZIMBABWE, ZW</u>.

O   Z   W   P.W   P.Z.

TOTAL : O   Z   W   P.W   P.Z

BIBLIOGRAPHY

THE FOLLOWING BOOKS HAVE BEEN HELPFUL IN THE PREPARATION
OF THIS LIST.   READERS WHO WISH TO PURSUE THEIR STUDY
WOULD DO WELL TO BECOME ACQUAINTED WITH THESE AUTHORS

Attenborough David, LIFE ON EARTH, Little,Brown and company, Boston,Toronto 1979
Bledsoe Thomas, BROWN BEAR SUMMER, Truman Tolleys books ,New York
Burt and Grossenheider, A FIELD GUIDE TO THE MAMMALS
Chanin Paul, THE NATURAL HISTORY OF OTTERS, Facts on file publications, N.Y.
Cousteau Jacques, MAMMALS IN THE SEA, the Danbury press 1975
Dagg Anne Innis and Bristol Foster , THE GIRAFFE, Van Nostrand Reinhold company
Ewer R.F., THE CARNIVORES, Cornell university press 1973
Frame Lory and George, SWIFT AND ENDURING, Elsevier Dutton co. 1981
Ghiglieri Michael , EAST TO THE MOUNTAINS OF THE MOON, Collier Mac Millan
N.Y.1988
Goodall Hugo and Jane, INNOCENT KILLERS, Mifflin company 1971
Grzimek's, ANIMAL LIFE ENCYCLOPEDIA (MAMMALS), 1975
Heintzelman Donalds, A WORLD GUIDE TO WHALES,DOLPHINS AND PORPOISES
Mac Donald David, THE ENCYCLOPEDIA OF MAMMALS, Facts on File N.Y.1988
McHugh Tom, THE TIME OF THE BUFFALO, Alfred A. Knopf inc. 1972
Mochi Ugo and Carter Donald, HOOFED MAMMALS OF THE WORLD, Charles
Scribner's, New York
Morris Desmond, PRIMATES ETHOLOGY, Aldine publishing co. Illinois 1967
Moss Cynthia, ELEPHANT MEMORIES, William Morron and company inc 1988
Much, THE WOLF, First university of Minnesota press 1981
Napier J.R.and P.H., THE NATURAL HISTORY OF THE PRIMATES
Neal Ernest, THE NATURAL HISTORY OF BADGERS, Facts on file inc. New York
Nokwak,Paradiso, WALKER'S MAMMALS OF THE WORLD
Perry Richard, THE WORLD OF THE POLAR BEAR, University of Washington press
1966
Perry Richard, THE WORLD OF THE WALRUS, Taplinger publishing New York
Sanderson I.T., LES MAMMIFERES VIVANTS DU MONDE
Schultz Adolph, THE LIFE OF PRIMATES, Universe books New York
Schwartz Jeffrey, THE RED APE,ORANGUTANS AND HUMAN ORIGINS, Houghton
Mifflin Co. Boston 1987
Sheldon G.William, THE WILDERNESS HOME OF THE GIANT PANDA, University of
Massachusetts press 1975
Strum Shirley, A JOURNEY INTO THE WORLD OF BABOON, Random house New York
1987
Taketazu Minoru, FOX FAMILY, Weather Hill / Heibonsha Tokyo 1979
Time life,   Wild Wild World of Animals 1977

INDEX OF SCIENTIFIC NAMES

INDEX OF SCIENTIFIC NAMES

INDEX OF SCIENTIFIC NAMES

351

INDEX OF SCIENTIFIC NAMES

INDEX OF SCIENTIFIC NAMES

INDEX OF SCIENTIFIC NAMES

355

INDEX OF SCIENTIFIC NAMES

INDEX OF COMMON NAMES

INDEX OF COMMON NAMES

INDEX OF COMMON NAMES

INDEX OF COMMON NAMES

INDEX OF COMMON NAMES

Guy Commeau was born in Paris, France. After 3 years in the French navy came to the U.S.A. in 1963.

He traveled extensively throughout the continental United States, Hawaii and Alaska. As naturalist and photographer his travels have taken him to Morocco, Algeria, Tunisia, Kenya, Japan, Hong-Kong, Taiwan, Malaysia, Singapore, Borneo, Mexico, Costa Rica, Peru, and most of the European Countries.

Commeau is President of the Redbud Audubon Society, elected and life member of the American Birding Association, member of the Los Angeles Zoo since 1977, Docent and Associate since 1985. Longtime member of other organizations, including the National Wildlife Federation, National Parks and Conservation Association. Guy Commeau and his wife Louise live in Clearlake Oaks, California.

For your free information.
To have your list published, send your Name and Address to:
          "Mammals and Countries of the World Association"
              P.O. Box 29, Clearlake Oaks, CA 95423

Additional copies of Mammals and Countries of the World:
A Check List, may be ordered by sending a check or money order for $24.95 each, plus $4.00 for postage and handling (paid by the publisher on orders of 2 books or more) to
Mammals and Countries of the World Association
P.O. Box 29, Clearlake Oaks, CA 95423

Requests for personal inscriptions by the author will be honored.